Social_Text_ **162**

Sound Carries: Coloniality, Race, and the Spatial Politics of Representation

Edited by Tariq Jazeel and Tom Western

Sound Carries

Techniques for Hearing Space and Politics

Tariq Jazeel and Tom Western

The contributions to this special issue stem from a workshop held in summer 2022, hosted by the Institute of Advanced Studies at University College London. Over a couple of days, we gathered friends and colleagues from a range of disciplinary backgrounds in the orbit of this institutional geography to share thoughts, ideas, and nascent work in progress, all of which had at its core a concern with the relationships among sound, space, and politics. Unsurprisingly, given our locatedness, in the process our temporary collective generated conversations that gravitated toward London— a London intractably positioned as postimperial metropolis, a city animated by ongoing colonialities, anticolonial echoes, and postcolonial potentials, and thus one whose postimperial spectral presence could not *not* haunt our conversations about many other elsewheres. We shared and heard work on radio infrastructures and wavelengths beyond empire, the musical geographies of independence and postcolonial nation building, the spatialities of British-Asian underground club nights in the metropole, the soundscapes of carceral spaces in the colonies, antiphonies and insurgencies between places sonically surveilled by the imperial state, and the multitudinous musics that cluster and swing together in the city and are always collapsing the distances between the privileged and excluded, or between empire's inner and outer spaces.

The geographies of the world, of course, are not neatly divided. There are many global Souths in the global North, and vice versa. Cartographies are jumbled in ways that confound the representational capacities of visual, two-dimensional maps. Our conversations—and this special issue—take as their point of departure the idea that there are maps that you can hear but not always see. London emerged as our staging ground

Social Text 162 · Vol. 43 No. 1 · March 2025
DOI 10.1215/01642472-11573328 © 2025 Duke University Press

for these spatial and historical sonorities. The city abounds with the sonic legacies of colonialism and imperialism—a city of sinister reverberations and silences, a city (like many others) formed through histories of imperialism.[1] As a result of these legacies, London is a whole world, meaning its sound cultures are a composition of intimacies and counterpoints, solidarities and syncopations, power geometries and tone rows, rhythms and repressions, counterrhythms and contestations, ballads and borders and hemispheric hemiolas.[2] The city itself is a set of worldly sonic geopoetics.

Although still so often framed as some kind of global center, our collective listening positioned London elsewise. Thinking in and with sound opens a way of hearing mobilities, movements, migrations, and relationalities that travel in many directions at once. This is true of sound in general. It is often stated that perceptions of sound—usually compared to vision—are multidirectional, rushing in from everywhere.[3] And here we want to extend this to thinking about sound and spatialities more broadly. Sound is not easily contained by the unidirectional colonial logics of centers and peripheries. Instead, it leaks and echoes and spills and constantly combines and recombines, becoming both an essential tool of presencing and placemaking—especially for marginalized communities—and a fundamental, even unremarkable, part of everyday life in diverse urban communities. In this respect, London is at once special and unexceptional. It is inevitably full of the entanglements produced by the imperial, postcolonial, and global world system, but so is everywhere else. As geographer Doreen Massey reminded us, all places are meeting places.[4] In sonic terms, then, the distinct rhythms and harmonics that make city life what it is are always a product of movements from many different histories and directions.

The articles in this special issue hear how sound *carries*—carries across geographies, histories, and disciplines; carries meanings, struggles, creative ways of being and knowing. It is our contention in what follows that, both politically and poetically, sound and space cannot easily be parsed.[5] Sound is spatial imagination and experience as much as it is social text. In this introductory article and in those that follow in this special issue, our concerted and collective effort is to stress the politico-intellectual value in thinking sound and space together in order to make sense of the worlds that people make together. We do so from a particular geographical context, with all the orientations and implications toward questions of coloniality and decoloniality that this brings, but in ways that seek some kind of fidelity to the unboundedness and dynamism of sound. We proceed from here by outlining some thoughts around cartographies that, we suggest, echo, before tuning in to some of the specific frequencies that constitute forms of world making and presencing in the diasporic city. We then frame the city as a "mixing desk" in order to foreground the political and spatial potential of thinking the city through sound, and we finish this introduction by outlining the multidisciplinary methodologies and approaches that move across the contributions that follow and that keep this special issue in time.

Cartographies That Echo

As we write this introduction, a year and a half on from our workshop, the city is periodically turned inside out.[6] Every couple of weeks, the center of London fills with people marching for ceasefire in Palestine and against genocide in Gaza. For an end to apartheid and ethnic cleansing. For liberatory futures. The city center becomes a sea of flags and voices, drawing connections between oppressed peoples and places and turning the metropolitan heart of empire into a giant anticolonial map. In the ephemeral space of the march—and in all the movement spaces that extend into the city beyond it—people are doing the work of reorganizing the world's histories and geographies to imagine life beyond imperialism. Sound helps bring this map into existence. The march moves through London like a planetary orchestra, with rhythms and chants assembled from many sources, at once being directed in anger toward the city's parliamentarians, in love toward the Palestinian people and Gaza, and in community toward the march itself and the ways it remakes the city and its own time-space extensions.

Sound has the potential to take on cartographic capacities, doing the work of opening spatialities beyond those drawn by compasses and quadrants (though plenty has also been made of the limits of sound in doing this work).[7] This is not to suggest a practice of plotting sonic distances and marking audible territories, but instead, as Ana María Ochoa Gautier writes, a process of making "an intellectual cartography that asks us to pay attention to the many possibilities of thinking in sound . . . —one that is less a map and more a conglomeration of stories."[8] This framing (also invoked by Sara Salem and Tom Western's contribution to this special issue) requires "listening across time and place in a manner that lives up to the challenges of twenty-first century geopolitics."[9] For our purposes, this means following sounds as they move, in constant feedback, bringing places closer together across physical distance and amplifying political and aesthetic commonalities that shuttle across histories and geographies.[10]

"In our thousands, in our millions." The Palestine march as a cartography of sound, story, and stretched spatiality does this connecting and amplifying. Sound carries to and from the march space, combining the cadences of chants from Arab revolutions, the second-line structures of brass band parades, the sound-system cultures of the Caribbean, the rhythmic pulse of the Brazilian *bateria*, the vocalizations of a solidarity movement that rings from city to city, and a set of new sonic practices rehearsed each time in efforts to keep the movement vital and keep the pressure on. In the process (and what we mean by the city being turned inside out), the usual veneers of an ever more tightly policed and unaffordable urban life are replaced by the "open song" of the colonized, marginalized, and precarious populations and those who stand in solidarity with them.[11]

The ways that sound carries, then, open ways of hearing geographies of relation and circulation. In the face of the static and vertically rooted culture-language area, Édouard Glissant gives us a lexicon of relay, poetics, and, importantly, echo that guides us here. Glissant's poetics traffics in the currency of art, language and literature, and sound and music to position the echo as a key mode of understanding, perceiving, and orienting ourselves to a world that always exceeds us but to which we are always already in relation. *Echos-monde*, as he calls them, are a series of places or people or ideas that are everywhere and simultaneously hold everywhere within themselves.[12] If this describes the anticolonial London gathered and galvanized in the space of the Palestine marches, it also helps us find ways out of the analytical tendency to position London as some kind of center, reproducing the colonial geographies of metropoles and outposts. Rather, thinking the city's polyphony through the lens of *echos-monde*, through Glissant's relational poetics, does the work of turning all peripheries into centers until the very idea of centers and peripheries is, in Glissant's words, abolished,[13] and in our reckoning, irrelevant.

Following this relational thinking more closely into worlds of sound and music, Louis Chude-Sokei writes of echo as "metaphoric of diversity and cross-cultural interaction," signaling both a production technique and a spatial relation.[14] Unlike the forced diversities and encounters produced through the imperial world system, however, echo brings things into relation "without the architecture of colonialism . . . to adjudicate or authorise hearing, meaning or blending."[15] The noun *echo* describes the process of a soundwave bouncing between surface and listener, the sound itself modulating in that movement. However, to experience an echo as a sound is to know that at some point the echo's source becomes irrelevant.[16] It is not a coincidence that echo is a technology and technique that carries from Caribbean philosophies and music cultures to the rest of the world, from the place where European racial taxonomies and hierarchies were so violently imposed and concretized. Yet within (and against) this violence, people found ways of "living, working, and surviving together."[17] Echo, for Chude-Sokei, is also a metaphor for reciprocity, making and holding space for community and connection beyond colonial forces. Echo bounces sound back and forth across diasporic geographies, creating new relations along the way.[18]

If sound carries like this, across distance, bringing people, and things, into relation, then it has the potential to precipitate and facilitate relations of care. This encompasses what Tina Campt calls "frequencies of care" when asking, "What would it mean to use sound to map the caring relations of our communities?"[19] As sound has long been a key means of evading, subverting, and resisting the colonial world forced onto people, it is also a sensory means by which people find and look after one another. A map made of frequencies of care resonates beneath the cartographies produced for empire, shaking and unmaking those imperial transparencies that always seek to keep people in place. Sound carries these relations and vibrations, these ways that people hold each other together and resonate with one another, across distance, irrespective of geopolitical power relations in the wake of empire.

What we are suggesting is that sound and music, waveforms and grooves, offer a creative means of finding paths out of the imperial legacies of mapping, classifying, cataloguing, authenticating, owning, dispossessing, oppressing. We do not mean to pose any simple binary opposition between the sonic and the visual here. Rather, we simply mean to contrapose the visual logic—the "way of seeing"—that inheres in the imperial map, codification, or category, to emphasize the cartographic possibilities of thinking space and sound together, of tracing the spatialities that sound and music produce, and of evoking the political potential and alternatives offered in Glissant's *echos-monde*. For Katherine McKittrick, the rhythms and echoes of diaspora create "a geographic praxis that is outside the colonial-imperial-capitalist logics," one that is beyond the logics of "market-time-nation-time-blood-and-land."[20] The Palestine march offers just such political, and poetic, potential for us. Its spatial promulgations sound out possibilities beyond our conventional maps. Its noise, its music, becomes experiment, invention, reparation, collaboration, a kind of collective freedom. And this in turn feeds into broader practices of diaspora geography, which "is not the act of making maps; rather, it is the act of sharing ideas about where liberation is and might be."[21]

Echoic cartographies thus do a kind suturing work. They are a set of bearings that people use to reassemble the worlds built by colonialism. They are "constellations of co-resistance,"[22] a set of stories, frequencies of care, freedom dreams, and inverted and inside-out geographies, all of which cannot be mapped beyond or outside the register of sound. Sound and what Jayna Brown refers to as "alter-frequencies" pose spatial/temporal folds in the here and now, ways "'out of the quagmire of the present.'"[23] From our workshop, and across the articles in this special issue, a vocabulary coalesces that facilitates this kind of listening: *amplification, antiphony, encryption, decryption, jamming, beatmapping, sonic modernities, subterranean movements, shadow histories.* This vocabulary works as a set of metaphors for analysis and critique as much as a set of sonic practices and production techniques. Together these ideas do cartographic work that opens spaces of audibility and ways of retuning the city.

World Making / Amplification / Presencing

Back to London. Another London:

> *di bredrin dem stan-up*
> *outside a Hip City,*
> *as usual, a look pretty;*
> *dem a lawf big lawf*
> *dem a talk dread talk*
> *dem a shuv an shuffle dem feet,*
> *soakin in di sweet musical beat.*
>
> *but when nite come*
> *policeman run dem dung;*
> *beat dem dung a grung,*
> *kick dem ass, sen dem pass justice*
> *to prison walls of gloom*
> (Johnson [1973] 2022)

The first verses of dub poet Linton Kwesi Johnson's 1973 poem "Yout Scene" paint a sonic picture. They describe a group of young Black British men gathering, chatting, laughing, listening, outside "Desmond's Hip City," a Brixton record shop serving this community with reggae music coming directly from Kingston Jamaica, and from London's Caribbean diaspora. The poem depicts an everyday scene, one of Black British conviviality, of a South London community whose Britishness is soaked in a "sweet musical beat" that makes distant and constitutive elsewheres audibly present.

As much as Johnson's words describe, they also demand to be read out loud. The patois in which "Yout Scene" is written precipitates its own enunciation, a kind of reading out loud that brings the verse to life. These are words whose textual and aesthetic force is embedded in the very sounds they command, sounds that carry the reader, or listener, to Johnson's London, a London whose cultural coordinates stretch to Jamaica and back, to Windrush and its colonial prehistory, to a racist postcolonial settlement in the heart of the empire in 1970s Britain, to a diaspora spatiality that cannot be represented *without* sound.

If Johnson's dub poetry speaks to the late twentieth-century British experience of inner-city Black youth, its transcendent effect is derived from its lyric and sonic qualities and capacities. Any oral culture makes its mark in the world through sound, the very trace that gives orality its intangible presence, its aesthetic form.[24] However, to stress how that orality makes a mark in the world through sound is to gloss the world-making capacities that sound possesses. We can be clear that, for the "bredrin" of "Yout Scene," the "sweet musical beat" is spatially, socially, and politically productive, just as the reading out loud of the poem is imaginatively and cartographically productive for any reader picking up a copy of this poem. In other words, sonic expressions and sonic cultures are forms of world making, what McKittrick calls "rebellious inventions."[25] Or, more precisely, sound helps us better understand forms of presencing that people use to make and sustain worlds. From broadcast infrastructures, through public listening practices, to the underground genres of fringe or radical music cultures, sound and music routinely produce spatiality and spatial lexicons for presencing.

"Outside a Hip City." That the collective listening in Johnson's poem is happening on the street is important. Amplified sound moves outward from the record shop and becomes public—a shared sonic iteration that makes (new) publics, that moves and is moving and that retunes the city.[26] In the context of postcolonial and Black histories, human migrations have precipitated struggles to come into representation in spatial and political contexts that routinely silence particular voices and forms of sonic expression. For marginalized communities in differential historical and geographical contexts, the challenge to speak and be heard has been key to staking claims on belonging. "Moving from silence into speech," as bell hooks puts it, "is for the oppressed, the colonised, the exploited, and those who stand and struggle side by side, a gesture of defiance that heals, that makes new life and new growth possible."[27] Amplification, both as sonic practice and as metaphor, is a political and a spatial praxis, coalescing into forms of presence and community. It is a form of world making.

Sitting in close proximity to Johnson's "Yout Scene," we can think here of the sound system culture that began to extend across London from the early 1960s. As Caspar Melville writes, it is not just that these elaborate and artisanal sound rigs moved across the city, providing London's Afro-diasporic communities with temporary, and mobile, nightclub and dance spaces in a context where licensed clubs, bars, and pubs were largely off-limits for Britain's Black population. It is also that the sound system was a multilayered, organized (and hierarchical) system—a business—involving

> the operator (who owned and managed and played the system live, as if it were a huge instrument), the engineer (responsible for the technical specifications and set-up for maximum volume and depth), the selector (who chose the records to play), the "DJ" (who vocalised over a microphone . . .), and the

box-boys (responsible for transporting and placing the amps and speakers and humping the heavy record boxes).[28]

Though patriarchal, hierarchical, and masculine, as Melville writes, the sound system provided an infrastructure for learning, training, and the generational transmission of extant skills in the community. For example, one technician in the Shaka sound system of south east London, Metro, had been an electrician in the army; another of the Sir Coxsone sound system worked on the air traffic control system at Heathrow Airport. The dances themselves, though not without their own power relations, were spaces of safety and care within Black communities whose access to the other public spaces in the city was still racially policed and curtailed.[29] In all these senses, sound system culture in London (and beyond) is just one example of how sound and music are world making, in and against the grain of the racially stratified, postimperial city.[30]

Ethnomusicologist Philip Bohlman lends a historical ear to this idea of a kind of world making, this amplification that, we suggest, effectively turns cities inside out. With a particular focus on religion, Bohlman traces European urban history through a convergence of religious soundscapes—Jewish, Christian, and Muslim—moving in overlap and counterpoint into the public spaces of cities across the continent. The sounds of worship, for Bohlman, "moved from the sanctuary to the public square, sometimes in gradual stages, but often through the dramatic modulation of public soundscapes." Jewish ḥazzanim (cantors) "turned outward to public spaces as liturgical music became cantorial music"; mosques moved from the courtyard to the street, with the adhān ringing out across city spaces.[31] Amplification was central to European religious and cultural diversity, "signifying the sonic limits of tolerance"[32] and, again, retuning cities.

Of course, the limits of tolerance—sonic and otherwise—in Europe are well known, and Bohlman also traces the attenuation and repression of these cultural counterpoints and diversities, both through European fascism and antisemitism and through racism and Islamophobia, where the effect of policing these alternative amplifications was to create "a public space of silence."[33] This in turn speaks back to wider colonial entanglements and the mutual constitution of sound and empire, at once on global scales and in ways that resonate through urban European histories. As Ronald Radano and Tejumola Olaniyan put it, "The emergence of European imperial orders and the concomitant rise of political democracies have also been matters of the ear."[34] Sound cultures, and sound technologies, have long been used to impose discipline and order across the colonial divide, both through militaristic violence and through the frequencies and wavelengths of imperial taxonomies. For example, the term *noise*, in particular, has a racialized and racializing history, being so often invoked as a marker of inferiority: "The command to silence, grew from an effort to contain the din—the noise of the 'Negro,' 'Chinaman,' and 'lazy native'—commonly portrayed in European travelogues over four centuries."[35] And European sonorities and systems of harmony continue to imagine their superiority, existing as the aural counterpart to Enlightenment notions of "reason" and "light."[36] We can hear in the modulation of city soundscapes a form of aural policing. In London, street musicians in the mid-nineteenth century were deemed a foreign presence in the city. These musicians—some of racialized European backgrounds, and others in the city owing to imperial migration—were grouped together as an "organ nuisance" in the press, labeled "Music-Grinders of the Metropolis" in the writings of the city's elites, and on the receiving end of noise legislation in the 1839 Metropolitan Police Act.[37] An 1860 cartoon in *Punch* magazine called for "a musical and political decontamination of English soil," with the visual depicting a group of musicians being booted off the cliffs of Dover into the Channel below.[38]

Into the twentieth century, the loud convivialities of urban life were subject to various noise-abatement campaigns and measures, often focused on regulating the use of sound playback technologies. As sound scholar Karin Bijsterveld recounts, efforts were repeatedly made to construct an essential difference between music played by musicians and music produced by playback devices, betraying a classist hostility toward sound cultures built on recorded music. The counterargument was that gramophones and radios—as relatively inexpensive sources of music—had become the musical instruments of the working classes, so to ban their noise would be to target these classes disproportionately.[39] Recorded music constituted a working-class sound culture, in which listening became an act of sharing music rather than making noise. In the 1950s, the Blaupunkt Blue Spot Radiogram, one of the earliest domestic record players, became a standard feature in West Indian homes in London, affording the ability to create a Caribbean soundscape at home in the diaspora.[40] As these technologies evolved into the sound system culture outlined above, Black institutions—often those associated with the sociabilities of sound—were regularly targeted by the police. For example, the Mangrove Café on All Saints Road on Notting Hill was raided twelve times in 1969–70, and the Metro Club, which regularly hosted sound system dances, was also frequently raided and targeted. They were not isolated cases.[41] Nor has the policing of Black noise gone away in London. In 2022, London's Westminster Council tried to ban people from gathering and playing dominoes in Maida Hill Market Square, in North West London, citing noise issues and antisocial behavior. The accused, a group of Black elders whose families came to Britain as part of the Windrush generation of the 1950s, took the council to court, claiming its order was racist and discriminated against Caribbean culture: "If you are West Indian, you just can't play dominoes without making a bit of noise."[42] The domino players won. The ban was lifted.

"*But when nite come / policeman run dem dung; / beat dem dung a grung, / kick dem ass, sen dem pass justice / to prison walls of gloom.*" For Linton Kwesi Johnson, and his depiction of Black British London of the early 1970s, Black life is always under threat from a whiteness that envelops, polices, incarcerates. Amplification, the sound that emanates from Hip City, is presencing; it is world making. But for that very reason, it can also be heard by power as a disruption. So when bell hooks goes on to assert that "speaking is not solely an expression of creative power; it is an act of resistance, a political gesture that challenges politics of domination, . . . it is a courageous act—as such, it represents a threat,"[43] we can expand this to say that alternative sound cultures more broadly often represent a threat to the imperial order, to the imperial state and its imperial interests, hence the policing of sound in London and across the colonized world, hence the efforts by the United Kingdom's Conservative government to paint the Palestine marches as extremist in late 2023/early 2024.[44] Just as sound carries across the colonial divide, so do efforts to police it as techniques of oppression tested in the colonies bounce back to the metropole.[45]

The ability to silence and suppress sonic expression, then, has been central to the exercise and operations of power in its many different manifestations.[46] As Johnson's "Yout Scene" reminds us, certain sounds, and orientations to sound, become indexed to ideas of race and ethnicity, and policing noise is a racial and carceral technology playing out along the sonic color line.[47] Music and sound, therefore, their articulation and manipulation, have been key vehicles and sites for decolonial, Black, and radical struggles. In other words, as Paul Gilroy writes, Johnson considers (sonic) culture "a vital force endowed with revolutionary potential, especially in the segregated worlds colonialism made."[48] Indeed, for Gilroy, Johnson's rebel disposition and the "freedom-seeking tradition to which he subscribes" are "nobly augmented by the poetic idiom he created."[49] These traditions and idioms, and their amplification, are always filling public space, expanding beyond the muffling techniques of imperial policing, retuning our cities.

Sounding the City / London as Mixing Desk

The Palestine march and Linton Kwesi Johnson's dub poetry locate us in London, which, as we have stressed, for this special issue is a geographical context central to the articles that follow. Johnson worked as a librarian at the Keskidee Centre on Caledonian Road in North London, and as is obvious from both his poetry and activism, Johnson's London was a "sophisticated anticolonial furnace" influenced by his involvement with the Caribbean Artist's Movement, the impact of Black Power, and his solidarity with the national liberation struggles that fed into the formation of the London-based Black Arts Movement.[50] Not dissimilarly, the Palestine march is a combination of sound and study. The words of poets and decolonial scholars abound on homemade placards; some people simply bring copies of Fanon's *Wretched of the Earth* to hold aloft. Many elsewheres congregate in the metropole, and the city itself becomes a sounding board for broader anticolonial geographical imaginations intent on working through the power geometries of overlapping territories and intertwined histories that colonialism has precipitated.

This has been the case for a while. Priyamvada Gopal writes of how London in the 1930s became an anticolonial "junction box," gathering oppositional figures and ideas from all across the British Empire. The irony is obvious, but it was in London that it was possible to articulate criticisms of British imperialism without the same degree of repression as in the colonies. And it was London, in Gopal's words, that "made it possible for colonial subjects—writers, intellectuals, labour activists, campaigners and journalists—to encounter each other, and to organise away from more repressive contexts."[51] The city became a space of "reverse tutelage," as anticolonial voices from colonized places at once internationalized British opposition to empire and sped up the push to end it altogether.[52] Throughout the twentieth century, London continued to be something of a relay, or maybe a switchboard, for decolonial soundings, studies, and spatialities that extend across the world.

These metaphors are appealing. However, if relays and switchboards are twentieth-century technologies of centralization and one-to-one communication, maybe it is more apt for us to imagine the city as a mixing desk: a multichannel, many-to-many model, with numerous inputs and outputs.[53] A mixing desk routes sound from many sources and then channels it outward for collective, public listening. Although the desk traditionally requires an engineer or DJ to do the virtuosic work of mixing and combining sounds, here we tap into a history of sonic arts that develops through the twentieth and into the twenty-first century and that moves away from individual(ist), composer-led approaches to sound technologies and Euclidean space—exemplified by the monumental compositions of Edgard Varèse in the 1950s—and toward collective, community practices in which everyone is at once listener, composer, and producer of sonic spatiality. This shift, traced beautifully by Gascia Ouzounian, creates a "confluence of acoustic, political, social, sensorial, and lived spaces."[54] And although we equally do not wish to fall too easily into the trap of automating the city, or imputing a computational model of urbanism,[55] the mixing desk thus presents a metaphor for combinatoriality, collaboration, and collectivity. The multichannel city is a space for the mixing of histories, geographies, and many different local knowledges, and its dynamic capabilities speak to a set of characteristics shared across sound production and movement spaces: of fades, volume and pitch control, time scaling, line outs, and channel inputs. The mixing desk, like the city, hybridizes sound, making something bigger than, or even just different from, the sum of its parts. It enables us to hear in many, many different ways, if only we make the effort to listen.

This special issue attempts this mixing work. It seeks to make space for, and articulate, some of the ideas outlined in this introductory article—cartographies that echo, world making, amplification and presencing, the city positioned as something like a mixing desk—in an attempt to contribute to a multi- and interdisciplinary decolonial sound studies that is steadily growing in volume. Each of the authors here, coming from different disciplinary perspectives, has an investment in sound, resulting in a collection of articles, conversations, and experimental approaches to working with and through music and sound that recognize their multiple relationships with, and in, the world. As an ephemeral collective whose conversations continue, we write variously from disciplinary backgrounds stretched across geography, sociology, history, gender and sexuality studies, music studies, and race studies.[56] By the same token, we all share some relationship with London. Yet this special issue is not *about* London in the sense of the city being a case study or compendium. Instead, and as we have suggested in these framing remarks, each article conglomerates narratives that collectively spill over neat notions of bounded and bordered space. Some are based in the city; for others it is background interference. The map, tapestry, or score that emerges here makes no claim to comprehensive geographical coverage. It is neither an atlas nor an anthology. Instead, this special issue offers a set of techniques that sound carries forward for thinking the relationships between politics and space.

Tao Leigh Goffe's article tracks the collapse, appropriation, and infrastructures of post-WWII radio broadcast technology, shuttling between London and the Caribbean in ways that bring into focus the postcolonial pulse of this intertwined spatiality and its dependence on soundwaves and the technologies that precipitated them. Her article teases out the textures, patinas, and trajectories of relational geographies forged between the Caribbean and the United Kingdom through music and sound, their technologies and infrastructures. Other articles in this special issue take London itself as the site of productive, if problematic, soundings whose political and historical implications for the production of ethnicized difference, emergent hybridities, and imaginative geographies are worth grappling with. Tariq Jazeel's contribution, which engages with the recent history of "Asian underground" music, also dubbed "New Asian Kool," stresses how this music's emergence in London's East End in the late 1990s was both enabled and constrained by its subterranean containment as a notionally "underground" form of cultural production. His article takes the underground not as a literal or normative countercultural space but instead as a metaphorical holding space for ethnicized difference, one produced by music and a "scene," and one that soon became replete with generic constraint. Its dissolution as a meaningful genre category, he shows, points to the umbilical relationship between British Asian (sonic) culture and the shifting topographies of late twentieth-century articulations of Britishness.

Likewise, Les Back and Stevie Back's article focuses on the work of contemporary London-based Black British musician Hak Baker, whose emergent singer-songwriter urban sound, and style, is dubbed by Baker as "guv'nor folk," or G-folk. Working closely with Baker and other folk musicians, including Angeline Morrison, Back and Back stress the importance of locating this music in a rich, antiracist tradition of British folk music (much of which is as urban as it is rural) that confounds any immediate association of British folk music with whiteness. Folk as music genre, they stress, must be thought and ontologically conceived as openly, relationally, and expansively as a city like London, and when it is, myths of purity and whiteness so associated with notions of folk themselves lose any necessary anchor or hold on the political imagination. Back and Back's narrative about the Black presence in British folk is sonified through a Spotify playlist that helps carry their argument.

In Sara Salem and Tom Western's article, which excavates anticolonial antiphonies across the Eastern Mediterranean, London would seem to give way to Cairo, Athens, and other elsewheres while continuing to lurk as a specter of colonial authority and sonic surveillance in the historical narratives that comprise their contribution. Their article adopts a nonlinear structure that ranges across geographies that defy colonial cartographies and classifications, bringing together places and anticolonial struggles that are so often understood separately, disjunctively, and serially. In teasing out the sonic resonances, echoes, and circularities between movements in Cairo, Athens, and beyond, Salem and Western bring into close connection an Eastern Mediterranean circuitry more often conceived in terms of a global geography of what Lisa Lowe calls "vast spatial distances."[57] From their looping narrative emerges an ongoing praxis of political resonance, carrying across time, space, and languages.

For a Sonic Chorography: Resonant and Rhythmic Methodologies

Taken together, the articles in this special issue constitute a kind of "sonic chorography," which is to say, a form of space writing that takes sound as its starting point. To be sure, sound is a kind of writing itself, but rather than thinking of this merely as phonography, we suggest *sonic chorography* to invoke the relations between sounding and spacing, or the relationships between sound and the (re)writing of place. This is our project here. From the electrical shops of postwar Jamaica, where pioneering engineers built loudspeakers that would change the world (Goffe) to the music nights in the East End of London that pulled British-Asian music scenes in and out of shape and South Asians into the national narrative (Jazeel), from the musical convivialities of the docklands a few miles farther southeast (Back and Back) to the sonic third worldisms that were broadcast over radios and chanted in the streets of Cairo and Athens (Salem and Western), this collection shows how sound is always making space and vice versa. By thinking chorography—literally "space writing"—as something audible, something aural, we take a neglected idea from the history of geography, sonify it, and hear how sound and space are in constant productive and political tension with each other.

And this, we argue, in closing this introductory article, amounts to a set of resonant and rhythmic methodologies for hearing the politics of spaces made and remade. They are *resonant* because resonance foregrounds the necessary open and generous relationality of sounding and listening. Borrowing from recent work in Indigenous sound studies, the relationship between the listener and listened to should not be one of subject-object but instead one of subject-subject. Resonance pushes against what Dylan Robinson refers to as a "western sense orientation" where we, as researchers, might simply "dismiss, affirm, or appropriate sound as content."[58] Our engagement with sound does not, in other words, conceive it simply as "data." Equally, we resist the easy liberal conflation of voice with subjecthood that inheres in modern society and is written into legal and political systems that make people legible before the state. (Think, for example, of the court or judicial "hearing" that imposes itself as the legalistic mechanism through which truth and/or justice can be administered.) Although we make no claims here of situating ourselves in, or contributing to, indigenous sound studies, nonetheless, by thinking the ways that sound carries, and what it produces in that movement, we are inspired by Robinson's methodological attunement to the "life, agency, and subjectivity of sound" within the context of the worlds from which it emerges.[59] This is what our analytical frequencies are attuned to in the articles that follow.

And these methodologies are *rhythmic* because this happens across both spatial and temporal registers. Against the Eurocolonial ordering of space and time (Greenwich Mean Time quite literally inscribes London as point zero of spatial and temporal subordination),[60] the sonic chorographies that unfold across these articles operate as a set of spatial counterrhythms. At the start of this article we suggested the term *hemispheric hemiolas* as a way of understanding the sound cultures that cohere in a city like London in the twenty-first century. The shifting cross-beats and counterrhythms of this musical-geographical figure open spaces of being and knowing in sound that carry across the territorialities of nations and borders, as well as the temporalities of pasts and futures. Rhythms, and counterrhythms, operate as resources and navigational tools for those of us seeking to hear and build worlds after empire. This special issue thus joins with recent work on sound, space, and politics: work that posits aurality as a necessary part of understanding colonialism;[61] work on audible infrastructures and intimacies that narrate musical imaginations and sonic geographies;[62] work on diasporic audio poetics, political reverberations, and sonic citizenships;[63] work that sounds out against colonialities built into structures of listening, organizing, and understanding sound and spatiopolitical relations.[64] Sound carries these conversations and, like rhythm, enjoins them to *keep on moving*. Sound carries life in our cities and keeps them reverberating. Sound carries this writing and the collectivities it engenders. Sound carries.

Tariq Jazeel teaches geography and race, ethnicity, and postcolonial studies at University College London. His research is positioned at the intersection of postcolonial theory, critical geography, and South Asian studies. He is professor of human geography and former codirector of University College London's Sarah Parker Remond Centre for the Study of Racism and Racialization, author of *Postcolonialism* (2019) and *Sacred Modernity* (2013), and coeditor of *Subaltern Geographies* (2019) and *Spatializing Politics* (2009).

Tom Western's work builds creative geographies that seek to imagine futures beyond the colonial past and present. He works primarily in Athens, where he is involved in various forms of creative and collaborative research and movement building. He is currently finishing his first book, titled *Circular Movements*, and has recently published pieces in *Society and Space*, *Journal of Creative Geography*, and *Migration and Society*. He is based at University College London as a lecturer in social and cultural geography.

Notes

We thank the Music Futures initiative at the University College London Institute of Advanced Studies for a grant that made the Sound Carries workshop possible in June 2022. In addition, we thank Sara Salem, Tao Leigh Goffe, Matthew Smith, Caroline Bressey, and Clive Nwonka for their participation in the workshop.

    1. Driver and Gilbert, *Imperial Cities*.
    2. A hemiola is a musical pattern whereby a group of two notes are played over

three beats, creating a two-three polyrhythm. Invoking this figure as a geographical idea evokes how the supposedly two separate halves of the world are always moving in the same times and spaces—albeit in ways that create offbeats and aporias, irresolutions created by colonialism that continue to be the backbeat of life in imperial metropolitan cities.

3. We are careful here not to essentialize or reify sound in any way or to reproduce what Jonathan Sterne calls "the audiovisual litany" of sensory perception in Western modernity, wherein the differences between hearing and seeing are often understood as a set of binaries: "Hearing is spherical, vision is directional; hearing immerses its subject, vision offers a perspective; sounds come to us, but vision travels to its object" (*Audible Past*, 14). As Sterne details, this litany is not a set of biological facts or some neutral description of the senses but comes into existence through a set of theologically informed ideologies that carries through the modern project. The audiovisual litany is loaded with value judgments, whereby the senses are organized and placed into the service of a politics that assigns certain tasks to vision (objectivity, clarity, reason) and to hearing (interiority, immersion, affect). Here we are particularly interested in collapsing the idea that hearing is primarily a temporal sense, with vision a correspondingly spatial one. Our thinking on this is also informed by the work of Gavin Steingo and Jim Sykes, who add to Sterne's litany the notion that "sound is Southern; vision is Northern" in their debunking of ideas that peoples and cultures in the global South are somehow "closer to sound and hearing than their European counterparts" ("Remapping Sound Studies in the Global South," 2–3). As with both of these projects, we avoid naturalizing and essentializing sound and pursue instead a relational, collective listening that follows the movements of sounds and their meanings across a set of entwined historical and geographical terrains.

4. Massey, *For Space*. On London's relational geographies specifically, see Massey, *World City*.

5. The connections between sound and space have been treated most extensively in Georgina Born's edited volume *Music, Sound, and Space* and in Gascia Ouzounian's *Stereophonica*. In her introduction, Born pulls together multiple bodies of scholarship to theorize sound and space across one another, including methods of spatialization that exist within music and sound art composition, in spaces of musical performance and sound installation, and in wider urban and cultural geographies. Particularly salient here is Born's reading that plenty of commonalities exist between spatial theory and sonic practice: both space and sound are inherently mobile and always in motion; both are intrinsically relational. In Born's words, music and sound "are particularly fertile conduits for spatial experience in that they have the capacity both to compound and to orchestrate in novel and affective ways the spatial affordances of social life writ large" ("Introduction," 24). Ouzounian, likewise, traces a history of thought and practice related to auditory spatiality as it emerges across such fields as philosophy, physics, physiology, psychology, music, architecture, and urban studies. Of direct relevance here is Ouzounian's narration of sonic propagation, or "the inherent ability of sound to pass through a medium and traverse space" (*Stereophonica*, 16). Tracking the historical developments of these ways of hearing, Ouzounian writes that "not only could sound propagate over long distances, its propagations could produce distinct effects—and not only sonic ones" (17). Other texts that feed into these ideas include Connell and Gibson, *Sound Tracks*; Feld, "Waterfalls of Song"; Jazeel, "World Is Sound?"; Kong, "Popular Music in Geographical Analysis"; Leyshon, Matless, and Revill, *Place of Music*; Saldanha, "Music, Space, Identity"; Smith, "Beyond Geography's Visible Worlds"; Stokes, *Ethnicity, Identity, and Music*; and Wood, Duffy, and Smith, "Art of Doing (Geographies of) Music."

6. Bayat, "Politics in the City-Inside-Out."

7. For a good historical account of this, see Ouzounian, *Stereophonica*, 125–50.

8. Ochoa Gautier, "Sonic Cartographies," 271. On audible territories, see LaBelle, *Acoustic Territories*.

9. Steingo and Sykes, "Remapping Sound Studies in the Global South," 24. Our project with this special issue in many ways builds on Steingo and Sykes's work with their volume *Remapping Sound Studies*. As the title suggests, Steingo and Sykes's volume also centers on sonic cartographies, albeit with more of an anthropological and ethnomusicological focus than this special issue of *Social Text*. Steingo and Sykes, and the work of the authors gathered in *Remapping Sound Studies*, do the important work of "listening to and from the South" (8) as a means of expanding and correcting the overwhelmingly global North focus of the sound studies canon as it has developed in Euro-American academia. Their efforts to "develop a new cartography for sound studies" (4)—and the navigational tools they craft to do so—enable us to hear the crossings and carrying work that sound does in constantly traversing the global North and South and traversing the colonial divides that produce Norths and Souths in the first place.

10. Again, Steingo and Sykes's writing is helpful here, as they conceive of sonic history as "a narrative of jagged histories of encounter, including friction, antagonism, surveillance, mitigation, navigation, negotiation, and nonlinear feedbacks, rather than as efficiency, inexhaustibility, increasing isolation of the listening subject, and increasing circulation" ("Remapping Sound Studies in the Global South," 12).

11. "Open song" here is a nod to Stefano Harney and Fred Moten, who write of "the open song of the ones who are supposed to be silent" (*Undercommons*, 51).

12. Glissant, *Poetics of Relation*, 93.

13. Glissant, *Poetics of Relation*, 29.

14. Chude-Sokei, *Sound of Culture*, 169.

15. Chude-Sokei, *Sound of Culture*, 169.

16. Ouzounian identifies echo as "that which can never be apprehended" (*Stereophonica*, 18).

17. Lowe, *Intimacies of Four Continents*, 34.

18. Our thinking here, and across this introduction, is also guided by Paul Gilroy's writing of *The Black Atlantic*.

19. Campt, "Frequencies of Care."

20. McKittrick, *Dear Science and Other Stories*, 182, 170.

21. McKittrick, *Dear Science and Other Stories*, 182.

22. Maynard and Simpson, *Rehearsals for Living*, 7.

23. José Esteban Muñoz, quoted in Brown, *Black Utopias*, 8.

24. Again, we do not wish to slide into a binary of orality and literacy here, which, for Sterne, parallels the spirit/letter distinction in Catholic religion and, leaning on Jacques Derrida, earmarks a "creeping Christian spiritualism that inhabits Western philosophy" (*Audible Past*, 17)—to the extent that this binary, and its latent religiosity, is essential to the history of the West itself. Our thinking here is closer again to Glissant, who asserts that, particularly in contexts of slavery and postslavery in the Black Atlantic, "the word is first and foremost sound" (*Caribbean Discourse*, 123–24).

25. McKittrick, *Dear Science and Other Stories*, 164.

26. For a good discussion of musical and sonic publics, see Born, "Introduction," 35–40. See also Hirschkind, *Ethical Soundscape*; and Ochoa Gautier, "Sonic Transculturation, Epistemologies of Purification, and the Aural Public Sphere in Latin America."

27. hooks, *Talking Back*, 9.

28. Melville, *It's a London Thing*, 60.

29. Melville, *It's a London Thing*, 58–62.

30. London's sound system culture, and its racialized policing, is depicted in Franco Rossi's 1980 film *Babylon*, which fictionalizes the story of Dennis Bovell's Sufferer's Hi-Fi in mid-1970s London. Steve McQueen's Small Axe film *House Party* (2020) is another depiction of London's sound system culture.

31. Bohlman, "Music Inside Out," 207–10.

32. Bohlman, "Music Inside Out," 213.

33. Bohlman, "Music Inside Out," 223.

34. Radano and Olaniyan, "Hearing Empire," 2.

35. Radano and Olaniyan, "Hearing Empire," 8.

36. Radano and Olaniyan, "Hearing Empire," 8; Erlmann, *Reason and Resonance*.

37. Picker, *Victorian Soundscapes*, 46–47.

38. Picker, *Victorian Soundscapes*, 48–49.

39. Bijsterveld, *Mechanical Sound*, 161–71. See also Mansell, *Age of Noise in Britain*.

40. Melville, *It's a London Thing*, 53.

41. Melville, *It's a London Thing*, 62.

42. Taylor, "Dominoes Player Wins Case Over 'Racist' Noise Ban in London Square."

43. hooks, *Talking Back*, 8.

44. Mureithi, "How the UK Government Rebranded Protest as Extremism."

45. Elliott-Cooper, *Black Resistance to British Policing*. Hervé Tchumkam, in "Banlieue Sounds," listens to similar dynamics in Paris, where the sonic-spatial specificities of the banlieues at the peripheries of the city, and how they are perceived as transgressions when they move into the spaces of the city center, are also a replication of colonial spatialities, policing, and order.

46. At the same time, we recognize that silence can also be a form of resistance and not just a key form of policing in the operation of colonial and state power (though it definitely is this). This is to say, the proclivity to silence particular voices—that is, to ensure some sounds, narratives, and expressions do not come into representation on terms true to the singularity of their own difference—is not the only meaning silence can have, either sonically or spatially. As Adrienne Rich writes in her poem "Cartographies of Silence" (*Dream of a Common Language*, 17),

> Silence can be a plan / rigorously executed / the blueprint to a life
> It is a presence / it has a history a form
> Do not confuse it / with any kind of absence.

Or, as Ochoa Gautier's "Silence" puts it, silence "can also be used as a significant political, symbolic, and interpretive strategy to respond to situations of conflict" (183–84). Far from existing as a binary of voice and silence, where one is agency and identity and the other is suppression and submission, these two sonic strategies merge into a praxis, a shifting set of textures, presences, and polyrhythms, both coming into representation and evading its demands.

47. Stoever, *Sonic Color Line*. George Revill also writes about sound as a tool of exclusion and inclusion in "Music and the Politics of Sound."

48. Gilroy, "Introduction," xix.

49. Gilroy, "Introduction," xvi.

50. See Gilroy, "Introduction," xvii.

51. Gopal, *Insurgent Empire*, 212.

52. Gopal, *Insurgent Empire*, 212–13, 279–318.

53. We are grateful to the journal editors for pushing us to think about the meanings and models of these metaphors.

54. Ouzounian, *Stereophonica*, 110.

55. See Mattern, *City Is Not a Computer*.

56. The symposium *Sound Carries* was part of an ongoing funded initiative at University College London's Institute of Advanced Studies called Music Futures, which grew from an experiment in bringing together scholars working on music (and sound in our case) in an institutional context that lacks a dedicated department of music. For more on Music Futures, see https://www.ucl.ac.uk/institute-of-advanced-studies/music-futures.

57. Lowe, *Intimacies of Four Continents*, 18.

58. Robinson, *Hungry Listening*, 15–16.

59. Robinson, *Hungry Listening*, 15.

60. Cheah, *What Is a World?*

61. Kanngieser, "Sonic Colonialities"; Ochoa Gautier, *Aurality*; Radano and Olaniyan, *Audible Empire*.

62. Born, *Music, Sound, and Space*; Devine and Boudreault-Fournier, *Audible Infrastructures*; Goffe, "Bigger than the Sound"; Hemmasi, *Tehrangeles Dreaming*; Hill, *Black Soundscapes White Stages*; Stokes, "Afterword."

63. Brent Turner, *Soundtrack to a Movement*; Brueck, Smith, and Verma, *Indian Sound Cultures*; Chávez, *Sounds of Crossing*; Denning, *Noise Uprising*; Weheliye, *Phonographies*; Western, "Covered Mouths Still Have Voices."

64. Goffe, "Unmapping the Caribbean"; Robinson, *Hungry Listening*.

References

Bayat, Asef. "Politics in the City-Inside-Out." *City and Society* 24, no. 2 (2012): 110–28.

Bijsterveld, Karin. *Mechanical Sound: Technology, Culture, and Public Problems of Noise in the Twentieth Century*. Cambridge, MA: MIT Press, 2008.

Bohlman, Philip. "Music Inside Out: Sounding Public Religion in a Post-secular Europe." In Born, *Music, Sound, and Space*, 205–23.

Born, Georgina. "Introduction—Music, Sound, and Space: Transformations of Public and Private Experience." In Born, *Music, Sound, and Space*, 1–69.

Born, Georgina, ed. *Music, Sound, and Space: Transformations of Public and Private Experience*. Cambridge: Cambridge University Press, 2013.

Brent Turner, Richard. *Soundtrack to a Movement: African American Islam, Jazz, and Black Internationalism*. New York: NYU Press, 2021.

Brown, Jayna. *Black Utopias: Speculative Life and the Music of Other Worlds*. Durham, NC: Duke University Press, 2021.

Brueck, Laura, Jacob Smith, and Neil Verma, eds. *Indian Sound Cultures, Indian Sound Citizenship*. Ann Arbor: University of Michigan Press, 2020.

Campt, Tina. "Frequencies of Care." Address presented at the Global Blackness Summer School, University of Johannesburg, November 26, 2022. https://www.youtube.com/watch?v=15I8OytxKrE.

Chávez, Alex. *Sounds of Crossing: Music, Migration, and the Aural Poetics of Huapango Arribeño*. Durham, NC: Duke University Press, 2017.

Cheah, Pheng. *What Is a World? On Postcolonial Literature as World Literature*. Durham, NC: Duke University Press, 2016.

Chude-Sokei, Louis. *The Sound of Culture: Diaspora and Black Technopoetics*. Middletown, CT: Wesleyan University Press, 2016.

Connell, John, and Chris Gibson. *Sound Tracks: Popular Music, Identity, and Place*. London: Routledge, 2003.

Denning, Michael. *Noise Uprising: The Audiopolitics of a World Musical Revolution*. London: Verso, 2015.

Devine, Kyle, and Alexandrine Boudreault-Fournier, eds. *Audible Infrastructures: Music, Sound, Media*. Oxford: Oxford University Press, 2021.

Driver, Felix, and David Gilbert, eds. *Imperial Cities: Landscape, Display, Identity*. Manchester: Manchester University Press, 1999.

Elliott-Cooper, Adam. *Black Resistance to British Policing*. Manchester: Manchester University Press, 2021.

Erlmann, Veit. *Reason and Resonance: A History of Modern Aurality*. Princeton, NJ: Princeton University Press, 2014.

Feld, Steven. "Waterfalls of Song: An Acoustemology of Place Resounding in Bosavi, Papua New Guinea." In *Senses of Place*, edited by Steven Feld and Keith Basso, 91–135. Santa Fe, NM: School of American Research Press, 1996.

Gilroy, Paul. *The Black Atlantic: Modernity and Double Consciousness*. London: Verso, 1993.

Gilroy, Paul. "Introduction: Culture as a Vital Force." In *Time Come: Selected Prose*, edited by Linton Kwesi Johnson, xv–xix. London: Picador, 2023.

Glissant, Édouard. *Caribbean Discourse: Selected Essays*. Translated by Michael Dash. Charlottesville: University Press of Virginia, 1989.

Glissant, Édouard. *The Poetics of Relation*. Translated by Betsy Wing. 1990; repr., Ann Arbor: University of Michigan Press, 1997.

Goffe, Tao Leigh. "Bigger than the Sound: The Jamaican Chinese Infrastructures of Reggae." *small axe* 24, no. 3 (2020): 97–127.

Goffe, Tao Leigh. "Unmapping the Caribbean: Toward a Digital Praxis of Archipelagic Sounding." *Archipelagos* 5 (2020): 1–23.

Gopal, Priyamvada. *Insurgent Empire: Anticolonial Resistance and British Dissent*. London: Verso, 2019.

Harney, Stefano, and Fred Moten. *The Undercommons: Fugitive Planning and Black Study*. Wivenhoe, UK: Minor Compositions, 2013.

Hemmasi, Farzaneh. *Tehrangeles Dreaming: Intimacy and Imagination in Southern California's Iranian Pop Music*. Durham, NC: Duke University Press, 2020.

Hill, Edwin C., Jr. *Black Soundscapes White Stages: The Meaning of Francophone Sound in the Black Atlantic*. Baltimore: Johns Hopkins University Press, 2013.

Hirschkind, Charles. *The Ethical Soundscape: Cassette Sermons and Islamic Counterpublics*. New York: Columbia University Press, 2006.

hooks, bell. *Talking Back: Thinking Feminist, Thinking Black*. New York: Routledge, 2015.

Jazeel, Tariq. "The World Is Sound? Geography, Musicology, and British-Asian Soundscapes." *Area* 37, no. 3 (2005): 233–41.

Johnson, Linton Kwesi. *Selected Poems*. 2002; repr., London: Penguin Random House, 2022.

Kanngieser, A. M. "Sonic Colonialities: Listening, Dispossession, and the (Re)Making of Anglo-European Nature." *Transactions of the Institute of British Geographers* 48, no. 4 (2023): 690–702.

Kong, Lily. "Popular Music in Geographical Analysis." *Progress in Human Geography* 19, no. 2 (1995): 183–98.

LaBelle, Brandon. *Acoustic Territories: Sound Culture and Everyday Life*. London: Continuum, 2010.

Leyshon, Andrew, David Matless, and George Revill, eds. *The Place of Music.* London: Guilford Press, 1998.

Lowe, Lisa. *The Intimacies of Four Continents.* Durham, NC: Duke University Press, 2015.

Mansell, James. *The Age of Noise in Britain: Hearing Modernity.* Urbana: University of Illinois Press, 2016.

Massey, Doreen. *For Space.* London: Sage, 2005.

Massey, Doreen. *World City.* Cambridge: Polity, 2007.

Mattern, Shannon. *A City Is Not a Computer: Other Urban Intelligences.* Princeton, NJ: Princeton University Press, 2021.

Maynard, Robyn, and Leanne Betasamosake Simpson. *Rehearsals for Living.* Chicago: Haymarket, 2022.

McKittrick, Katherine. *Dear Science and Other Stories.* Durham, NC: Duke University Press, 2021.

Melville, Casper. *It's a London Thing: How Rare Groove, Acid House, and Jungle Remapped the City.* Manchester: Manchester University Press, 2020.

Mureithi, Anita. "How the UK Government Rebranded Extremism as Protest." *Open Democracy*, March 11, 2024. https://www.opendemocracy.net/en/uk-palestine-protesters-extremism-islamophobia-gove-makram-ali/.

Ochoa Gautier, Ana María. *Aurality: Listening and Knowledge in Nineteenth-Century Colombia.* Durham, NC: Duke University Press, 2014.

Ochoa Gautier, Ana María. "Silence." In *Keywords in Sound*, edited by David Novak and Matt Sakakeeny, 183–92. Durham, NC: Duke University Press, 2015.

Ochoa Gautier, Ana María. "Sonic Cartographies." In Steingo and Sykes, *Remapping Sound Studies*, 261–74.

Ochoa Gautier, Ana María. "Sonic Transculturation, Epistemologies of Purification and the Aural Public Sphere in Latin America." *Social Identities* 12, no. 6 (2006): 803–25.

Ouzounian, Gascia. *Stereophonica: Sound and Space in Science, Technology, and the Arts.* Cambridge, MA: MIT Press, 2020.

Picker, John. *Victorian Soundscapes.* Oxford: Oxford University Press, 2003.

Radano, Ronald, and Tejumola Olaniyan, eds. *Audible Empire: Music, Global Politics, Critique.* Durham, NC: Duke University Press, 2016.

Radano, Ronald, and Tejumola Olaniyan. "Hearing Empire." In Radano and Olaniyan, *Audible Empire*, 1–22.

Revill, George. "Music and the Politics of Sound: Nationalism, Citizenship, and Auditory Space." *Environment and Planning D: Society and Space* 18, no. 5 (2000): 597–613.

Rich, Adrienne. *The Dream of a Common Language.* New York: Norton, 1978.

Robinson, Dylan. *Hungry Listening: Resonant Theory for Indigenous Sound Studies.* Minneapolis: University of Minnesota Press, 2020.

Saldanha, A. "Music, Space, Identity: Geographies of Youth Culture in Bangalore." *Cultural Studies* 16 (2002): 337–50.

Smith, Susan J. "Beyond Geography's Visible Worlds: A Cultural Politics of Music." *Progress in Human Geography* 21, no. 4 (1997): 502–29.

Steingo, Gavin, and Jim Sykes, eds. *Remapping Sound Studies.* Durham, NC: Duke University Press, 2019.

Steingo, Gavin, and Jim Sykes. "Remapping Sound Studies in the Global South." In Steingo and Sykes, *Remapping Sound Studies*, 1–36.

Sterne, Jonathan. *The Audible Past: Cultural Origins of Sound Reproduction.* Durham, NC: Duke University Press, 2003.

Stoever, Jennifer. *The Sonic Color Line: Race and the Cultural Politics of Listening.* New York: NYU Press, 2016.

Stokes, Martin. "Afterword: A Worldly Musicology?" In *The Cambridge History of World Music*, edited by Philip Bohlman, 826–42. Cambridge: Cambridge University Press, 2013.

Stokes, Martin. *Ethnicity, Identity, and Music: The Musical Construction of Place.* Oxford: Berg, 1994.

Taylor, Harry. "Dominoes Player Wins Case Over 'Racist' Noise Ban in London Square." *Guardian*, May 14, 2022. https://www.theguardian.com/uk-news/2022/may/14/dominoes-player-wins-case-over-racist-noise-ban-in-london-square.

Tchumkam, Hervé. "Banlieue Sounds, or, The Right to Exist." In Steingo and Sykes, *Remapping Sound Studies*, 185–202.

Weheliye, Alexander. *Phonographies: Grooves in Sonic Afro-Modernity.* Durham, NC: Duke University Press, 2005.

Western, Tom. "Covered Mouths Still Have Voices." *Journal of Sonic Studies* 24 (2023): 1–20.

Wood, Nichola, Michelle Duffy, and Susan J. Smith. "The Art of Doing (Geographies of) Music." *Environment and Planning D: Society and Space* 25, no. 5 (2007): 867–89.

Anticolonial Antiphonies

Sara Salem and Tom Western

Antiphony is a vocal practice. Usually meaning call and response, and part of many forms of music, antiphony sounds community and makes community in sound. In Paul Gilroy's formulation of the Black Atlantic, antiphony bridges from music to other forms of cultural expression, doing the work of connecting and relating among people that live across seas and oceans yet are bound together in displacement and diaspora. There is, for Gilroy, "a democratic, communitarian moment enshrined in the practice of antiphony which symbolises and anticipates (but does not guarantee) new, non-dominating social relationships."[1] Antiphony is equally a protest sound. In demonstrations and occupations, antiphony is the back-and-forth of chant, the consensus making and collective demand of the people's microphone. It is a Greek word, αντιφωνία, that in Greek means something different again, something closer to contradiction or discord.

This is an article about antiphony. In particular, we seek to narrate anticolonial antiphonies, in which resonate all of the qualities above: protest, connection, and contradiction. In anticolonial antiphonies we hear an ongoing praxis of political reverberation and relay, a shuttling of strategies that carry across time, space, and language, connecting decolonial movements across histories and geographies. Neither antiphony nor anticolonialism is a smooth process—not a simple echo, repeating and reproducing politics at a distance. Instead, antiphony inserts discord and dissonance into colonial orders, seeking to bring other worlds into being. By putting anticolonialism and antiphony together, we hear how sound carries movements across geographies that defy colonial cartographies and classifications, bringing together places and struggles that are often understood separately.[2] These movements speak to one another, backward and forward through history, producing unruly tempos that push against the linear logics of colonial time.

Social Text 162 · Vol. 43 No. 1 · March 2025
DOI 10.1215/01642472-11573341   © 2025 Duke University Press

Following these antiphonal trajectories, we have written this article without a linear structure. The antiphonies we narrate here follow geographies, histories, and movements that move in multiple directions. We seek to reflect that in our writing. We do so by combining our work on anticolonial histories, afterlives, and futures in Cairo, Egypt, and Athens, Greece, and in the process we hear the Mediterranean—particularly the Eastern Mediterranean—as a space that holds multiple imaginations of decolonization and futurity.[3] The article tells four stories, moving antiphonally between these two cities, which become a single narrative that jumps between scenes and sounds, times and places.[4] The article itself, then, is a set of anticolonial antiphonies: a series of stories (let's call them "Mediterr-ations") that loop into, across, and sometimes against one another and that don't begin and end in any orderly manner. We begin by placing anticolonial antiphonies into bigger histories of third worldism, before listening to revolutionary and counterrevolutionary soundscapes, "beatmapping" practices that connect migration and mobilization, broadcast voices that affectively disseminated anticolonial resistance, and the insurgent geographies created by sonic uprising. We end with a brief and speculative conclusion, as we imagine these stories continuing to loop into one another and joining with other anticolonial antiphonies across time and place.

Sonic Third Worldism (from City to City and Shore to Shore)

In an essay on the 1955 Asia-Africa Conference in Bandung, Indonesia, Khadija El Alaoui writes this event as not just a gathering of political leaders, setting to the task of reorganizing the world's histories and geographies at the end of empire. It was also an event that lived and was carried in the streets of colonized places around the world, in the rhyme and rhythm of people's poets, in the voices of everyday conversations. "Street Bandung," for El Alaoui, is "where neighbourliness is practiced," where "new ways of doing justice are imagined."[5] Street Bandung is a vernacular political composition, voiced polyphonically and resounding across continents, connecting anticolonial struggles along the way.

Across the world, and at the same time, people from Cairo to Algiers were buying radios and batteries to tune in to newly broadcast stations such as Voice of the Arabs and Voice of Algeria, evading and combatting imperialism with the weapons of wavelengths and radiophonic relay. In Frantz Fanon's famous account, radio listening was the means through which the Algerian people felt themselves "to be called upon and wanted to become a reverberating element of the vast network of meanings born of the liberating combat."[6] Broadcast sound was a central means of combating colonial occupation and "believing in the liberation."[7] Across the world again, artists and writers in the Caribbean were simultaneously experimenting with forms of sounding that could offer countertestimony to colonial histories and open new poetics of diasporic relating. Sonic modes of representation functioned as "a sharp (temporal and spatial) point of exploration, making resonant a history of loss."[8]

Street Bandung, anticolonial radio, Négritude audiopoetics—in all these cases (and countless others) anticolonialism was constituted in and through sound. We hear this as a kind of sonic third worldism, which at once contains all of the stories that follow in this article and places the sonic expressions that we analyze into a much bigger political history. Third worldism itself emerges from the anticolonial struggles of the mid-twentieth century, sometimes narrated as being bound between 1945 and 1989, or what Michael Denning calls "the age of three worlds"; sometimes it is placed into longer histories of contestation against empire stretching back to the beginning of European empire in 1492 and into the present.[9] The third world was not a place but a project, and here our task of hearing a sonic third worldism is twofold.

First, it requires tuning in to the echoes of third worldism that continue to reverberate in the present. Colonialism itself hasn't ended; never has this been clearer than in the present we are writing in, when a whole world calls for an end to settler colonialism and the genocidal violence that extends from it, and another world carries on regardless. This means that anticolonialism, too, is an ongoing practice, constantly seeking to connect peoples and places and build toward collective forms of liberation. Third worldism is thus reactivated in the political present. We argue that sound carries these anticolonial and liberatory politics in ways that collapse far-flung geographies and neat distinctions between past and present. Half of the stories in this article take place after the age of three worlds had supposedly come to an end, but they are bound together through antiphonies that jump across time and place.[10]

Second, it involves foregrounding the geopoetics of third worldism as much as its geopolitics. This was a movement of cultural workers and popular struggles as well as of statesmen and intellectuals. Recent work on cultures of anticolonialism highlights the creative labor of writers' conferences and resistance literatures, the poetic solidarities of decolonial experiments and new aesthetic imaginaries, and the role of recorded music in circulating noisy alternatives to empire.[11] Our hearing of sonic third worldism, then, joins with scholarship that engages the sonic politics of anticolonialism.[12] And it joins equally with the work of sound artists and musicians who plug in to these histories and bring them resounding into the present.[13] So although our focus here is on a particular set of times and places—Cairo and Athens, from the middle of the last century to the present—the sounds we follow and the sonic strategies we narrate are always speaking to something much bigger: a sonic third worldism that carries through antiphony between these two cities and onward to many other places. Numerous forms of colonialism and imperialism are entangled in these stories, just as multiple forms of resistance are connected in efforts to make worlds beyond them.

Revolutionary and Counterrevolutionary Soundscapes

The 2011 Egyptian revolution was a pivotal political and social event, an eighteen-day struggle against a military dictatorship led by Hosni Mubarak that had been in power since 1981. It was also a visceral event during which new modes of politics were produced and reproduced. We are interested here in the viscerality of this event: the way bodies and emotions flowed through the public spaces in which revolutionary resistance was taking place. In *The Visceral Logics of Decolonization*, Neetu Khanna centers the visceral nature of anticolonialism and revolution more broadly, arguing that political events did not simply provoke feelings but were also produced in and through visceral and embodied states.[14] The body in revolution is a body creating revolution, and here the sonic viscerality is as important as other forms of visceral feeling. The soundscape of the 2011 revolution was not merely a backdrop but a central dimension of revolutionary politics.

The revolution erupted after a decade of strikes and protests reverberated around Egypt calling for social justice and democracy. As news that Hosni Mubarak planned to hand over power to his son Gamal began to spread, Egyptians began to mobilize in favor of holding democratic elections. This built on mobilizations calling for social justice, targeted primarily at the wave of privatizations sweeping the public sector, the rising cost of living, and rising inequality, as well as the growing unrest around police brutality that was seen as expanding into the everyday lives of most Egyptians. The revolution itself, which began on January 25, 2011, was inspired by the revolution that broke out in Tunisia in December 2010. Indeed, one of the main chants heard in Tunisia, *Al sha'ab yureed esqat al nizam* (The people demand the fall of the regime), was to become one of the most common chants in Egypt.

In this section, we explore the politics of sound through the various sounds that make up a revolution. Sound was not a consequence of political action but instead did political work; it was mobilized and engaged as a means of coming together while simultaneously documenting forms of violence against protesters. As Elliott Colla writes, "Poetry is not an ornament to the [Egyptian] uprising—it is its soundtrack and also composes a significant part of the action itself."[15] Scholarly work has focused on the role of poetry and chanting during the revolution as a particularly prominent aspect of political mobilization.[16] The centrality of music and poetry has a longer history, dating back to figures like Ahmed Fuad Negm, whose poems (with music from Sheikh Imam) were central to resistance against the government during the 1970s, and whose poetry was revived during 2011.[17] Throughout the revolution, particular figures became associated with revolutionary songs. Ramy Essam, for instance, wrote a song titled "Irhal" that was played over and over in Tahrir Square.[18] This has led scholars to understand the poet as being at the forefront of the revolution.[19]

Sounds such as chants, songs, poems, music, police guns, fireworks, tear gas, sirens, and food vendors make up a soundscape of revolution that acts as an important archive on the viscerality of protest. As Khanna notes, this type of viscerality produces its own tempo—as opposed to a linear or rationalist historical teleology, the visceral tends to produce a disruptive and erratic temporality:[20]

> The visceral inhabits an unstable time full of potentiality, of possibility and destruction, the erratic and volatile time of the colonised and the subaltern. The time of the visceral disrupts the linear, homogenous, empty time of bourgeois historicism; it refuses the "spatialized, measurable, quantifiable, homogenous, empty, and teleological time" of capitalist modernity and bourgeois historicism and its sequencing of "always fleeting and inconsequential...instants" where each gains significance only through its own negation.[21]

Revolutionary sounds similarly did the political work of altering the experience of time, producing tempos that were disruptive and erratic rather than linear and under control. These came up against counterrevolutionary sounds, such as police guns, police sirens, and tear gas cannisters going off, which were similarly doing their own political work of provoking chaos, fear, and paranoia. The soundscapes of revolution and counterrevolution are thus not aftereffects of politics but constitutive of how politics unfolds.

Take, for instance, the soundscape of the prison, which we can think of as a space of counterrevolution. In "The Sound of Prison," Lina Attalah sketches out the lifeworld of Tora, one of Egypt's most well-known prisons.[22] The visuality and soundscape evoked through the piece bring to life the prison space beyond what we might imagine when we think of prisons, rendering them spaces of both life and death, politics and despair. Watching as families enter the prison to visit loved ones, she notes in passing that the scene "feels like an invisible face of Cairo."[23] The prison has its own economy—vendors selling oranges and mandarins, trays of *konafa* (a Middle Eastern pastry soaked in syrup), and clothes. The sounds of Cairo blend with the sounds of prison life, the cacophony of conversations and arguments making up a "temporary prison soundscape."[24]

The piece is striking in its affective sketch of a range of experiences: the fatigue of families coming to visit, the impatience of detainees waiting for them to arrive, the search for privacy once they are reunited. The sonic landscape of revolution thus also includes the sounds of what has been a brutal counterrevolution, of which the prison has been a central mechanism of repression. Similarly, Lawrence Abu Hamdan's 2016 work "Saydnaya (the Missing 19dB)," focusing on Saydnaya Military Prison, thirty kilometers north of Damascus, Syria, explores the soundscape of prison.[25] Using memories of former detainees that are largely sonic, Abu Hamdan explores how "sound and silence are connected to techniques of domination, power and resistance" through "forensic listening."[26] Silence emerges as the dominant feature of the prison soundscape, a silence that is felt as part of the brutality of the prison itself.[27]

Thinking of revolutionary chants as anticolonial antiphonies foregrounds the sonic elements of resistance while also placing the revolution in a longer trajectory of anticolonial struggle—against a postcolonial state invested in reproducing colonial dynamics.[28] Revolutionary chants connected geographies across the Middle East and North Africa, created public collectives in protest spaces, issued warnings about police violence, and structured movement during protests. This latter point in particular highlights the way chants were constitutive of revolt, as opposed to merely producing a soundtrack of it. As Egyptian historian Alia Mossallam recalls,

> *Ithbat.* Ithbat can be translated as "stand still," "steady" and "unshaken." It was one of the newer chants that were infused into us on the 25th of January 2011—every time the police launched an offensive, and people started to run, someone would shout "Ithbat" as he or she stopped moving, and then several would shout it, and then tens and hundreds, until thousands would stop. I would close my ears and squeeze my eyes shut and let the thousands of voices shake through me, shake out the fear, and stabilise my resolve.[29]

In this sonic memory, Mossallam draws out not only how the chant contained a political call to stand still and resist but also that the thousands of voices chanting would "shake out the fear," allowing her to remain in place. As Colla notes, the sense of collectivity created through chanting together allowed protesters to let go of their fear.[30] Chants similarly connected protests across Egypt to one another through sound as chants emerging in Suez might reverberate to Cairo and chants from village squares echoed through big cities. Even when sound was not traveling directly, it was doing the work of connecting revolutionary demands through visceral and embodied experiences. Chants also allowed for revolutionary spontaneity, as protesters came up with rhyming couplets on the spot as a means of capturing their feelings.[31]

Expanding on the power of revolutionary chants, Mossallam describes how they allowed people brought together collectively to decide where to go and what to do. If a chant "captured the imagination," then it would inspire people to take action; if it didn't, it would disappear quietly into the crowd. Chants also travel transnationally; they are forms of moving sound. One of the most popular chants in Egypt came from Tunisia: "The people demand the fall of the regime." This chant traveled across North Africa and Southwest Asia, creating a soundscape of regional revolution. Beginning in Tunisia, it moved to Egypt, then Syria, then Saudi Arabia, then Yemen, and beyond. The chant is in classical Arabic, which is distinctive from most of the other chants, which in turn explains how it spread across geographies.[32] This chant came to define these different geographical spaces through a collective set of political demands, more so than other outputs such as social media posts or announcements by revolutionary groups. Here the form of the chant—specifically its linguistic form—creates internationalist connections. The intonation, accent, or specific words of the chant might change as it shifts location, but its form and rhythm remain the same. In this sense, sound became integral to creating a shared political space.

In contrast to spontaneity, chants also often involved historical repetition, a characteristic that speaks to the form chants take. Poetry, songs, and chants from previous uprisings became part of the present through repetition, and the form of the chant allowed for this because of the familiarity of these past sonic echoes. This familiarity is linked to its rhythm and memories this brings up, rather than its content. As such, familiarity and spontaneity both came to define revolutionary chants. Chants mobilize people, often through reference to the past, while creating an opening in the present to imagine a different future. As Mossallam writes, "A chant can move and mobilise like poetry, it can capture your heart like a song, and it can weave a future before you like a vision."[33] Mossallam beautifully describes the viscerality that Khanna alludes to in her work, showing how it is not simply a background sensation but an embodied experience producing political change. Invoking affect, it speaks to the power embedded within sound, to connect across geographies and temporalities while at the same time anchoring the listener firmly in the present. Soundscapes of revolution and counterrevolution therefore have much to tell us about Egypt in 2011, as well as how resistance and the sonic are entangled. Similarly, chants connected to past revolutions created connections with the 2011 revolution. The cartographic work being done here disrupts fixed spatial boundaries and linear temporalities at the same time.

Beatmapping the Mediterranean

It's 2021, a summer evening in Athens. Members of a youth organization, founded and led by people of refugee background, form a circle in front of the Greek national academy and prepare to perform *arada*, the traditional performance art of Damascus, Syria, made out of rhyme and rhythm, song and pun, call and response. Two *daouli* and a *darbuka* sit on the pavement, both percussion instruments with long migratory histories in aural public spheres all around the Eastern Mediterranean.[34] A crowd assembles in front of the academy building, designed to resemble imaginings of the ancient city but actually built in the nineteenth century by Danish architects with Austrian financing. The air fills quickly. Chants resounding with Damascene revolt against empires—against Ottoman imperialism and the French mandate—ring out into Athenian public space.

Ten years earlier, these same chants provided the 2011 uprising in Syria with its revolutionary rhythms, where people in Syrian cities joined the chorus of rebellion across the region, vocalizing and amplifying the same appeal that carried from Tunisia and Egypt—the people demand the fall of the regime—while also creating new chants built onto traditional *arada* forms.[35] These chants combine celebration with subversion, speaking historically to the honor and courage of people seeking to keep their city free from injustice and the people free from oppression—folding historical anticolonialisms against French and Ottoman rule into revolutionary demands in the present.

In their updated twenty-first-century form, the chants of *arada* were "some of the most creative cultural productions of the revolution."[36] Every Friday people would assemble in cities across Syria, gathering around a particular theme that gave impetus to the uprising, with local committees coordinating slogans to unite the movement.[37] Chants did unifying work, antisectarian work, geographical work. Voices shouted solidarity with places that were being besieged and attacked by the Assad regime forces. Holding everything together was the collective vocalization, "One, one, one, the Syrian people are one."[38] Chants carried from city to city—from Damascus to Homs to Deraa to Aleppo to Idlib and so on—piecing people together in the process, and collapsing the techniques of distancing and division that had underwritten the regime over the previous four brutal decades.[39] While the chants in 2011 and after were not anticolonial in any straightforward sense, as was the case with neighboring revolutions, they propelled the movement against dictatorships that had prevented the full realization of anticolonial projects in the region, linking uprising to previous moments of anticolonial resistance and voicing them all at once.

In Athens, ten years later, the *arada* performance contains shoutouts to Syria and Damascus, to Palestine and to Greece and Athens itself. The revolutionary practice of cities singing for other cities is stretched and echoed across the Mediterranean.⁴⁰ These practices now speak of both migration and mobilization. Athens—and other places along with it—is mapped into geographies of liberation. As Robin Yassin-Kassab and Leila Al-Shami note, creating a new geography of liberation was both a guiding force and a consequence of the uprising in Syria, one that did not map onto to the old cartographies of empire.⁴¹ Instead, the chants of revolution build an internationalism on its own terms. This is similar to what Ilham Khuri-Makdisi calls a "popular anti-imperialism" when detailing geographies of contestation in the region in the late nineteenth century.⁴² In that moment, political ideas bounced between urban spaces around the Eastern Mediterranean, fostering a sense of global radicalism as people organized against direct and indirect forms of imperialism, which in Athens now becomes a transhistorical and multipolar anticolonialism, spoken out against Ottoman and French imperialisms in the past, against the colonialities of EU border politics that structure and limit life in the present, and against forms of oppression and intervention in between—layering them all together in rhyme and verse. People continue to build geographies and futures from the ground up, futures that are plural and participatory and that hold people and places together. "We will sing for the revolution that we believe in," Hani Al-Sawah asserts in the essential volume *Syria Speaks*, "for the cities we love and for the freedom that all of us seek."⁴³

We hear this as an anticolonial antiphony—call of contestation, response of reaffirmation—and as a kind of "beatmapping." In music production, beatmapping is a tool for detecting and developing rhythms across tempos and keeping beats in propulsion. We reframe beatmapping as a cartographic practice. Calling out city names and affirming each other's uprising is a way that people find one another in struggle and resistance, which in itself is an act of movement building and world making. Beatmapping, in this sense, brings movements and places into relation. The sound of drums and voices calling out for one another makes maps that moved beyond the logics of sectarianism, division, and oppression in Syria and that move beyond the colonialities of European borders and the ways they separate people, continents, and movements.⁴⁴ Rather than reproducing these border logics, and the rhythms of pushbacks and narratives of the sea as a graveyard, beatmapping turns the Mediterranean into a space of feedback and contestation.

The street where the *arada* performance takes place is one of the big boulevards in Athens: Panepistimiou, or University Street, a road written in broad strokes when Athens was installed as capital of newly independent Greece by the "protecting powers" of the 1820s. With political and urban planning from London, Paris, and Moscow, the city was built in (and as) a European vision. Yet beneath (and against) this top-down

Europolitical culture a street poetics has always existed that positions Athens as a Mediterranean city, what urban theorist Ioanna Theocharopoulou calls the "popular-Eastern Mediterranean."[45] This is something that plays out in sound—in language and music, in rhythms and uses of urban space—and affirms sonically and culturally what maps already tell us: Athens is closer to Cairo, Damascus, Beirut, and Jerusalem than it is to London, Paris, and European "protection." On this hearing, the chants of *arada* in the center of Athens make spatialities that are at once new and very old, amplifying entangled Mediterranean histories and turning the city around to face the sea.

The anticolonial resonances of the chants themselves are significant. Vocalizations against historical imperialisms bring the past beating into the present, galvanizing and strengthening contemporary movements through an accumulation of resistance(s). Anticolonial histories are recuperated into revolutionary presents. Again this disrupts the tempos of linear time, which is also colonial time. As Carolyn Nakamura writes, European colonization involved the colonization of time and the replacing of the plurality of histories with the collective singular History.[46] But histories reside in bodies, which make revolution out of history. Chants sound out polyrhythmic pasts that crescendo in the present, and they become a gathering space for expanded solidarities. In 2011, Syrian people chanted to support revolution in Egypt.[47] From Athens in 2019, displaced Syrians chanted to support revolution in Sudan.[48] This beatmapping traces the contours of struggle and solidarity, pulsing back and forth across cities and seas.

These sung politics make a geography that is also not linear, that moves in multiple directions. The *arada* sung to Athens, to Cairo, to Damascus, to Khartoum, to Palestine holds movements together. Anticolonial antiphonies do both a carrying work and a cartographic work. They do the work of "inventing places from which to speak,"[49] as people continue to find creative ways to resist the compound forces of dictatorship and counterrevolution, the border and the refugee camp. The rhythms of *daouli* and *darbuka*, of chant and song, of histories and geographies, of revolution and uprising become spaces of invention and of countercartography. As geographer Nour Joudah puts it, "Maps can confine and erase peoples and places, but they can also free a vision. Spatial imagination provides opportunities that policy debates do not."[50] Here this is a spatial imagination made audible, what Stefano Harney and Fred Moten call "the joyful noise of the scattered."[51]

These attunements are not easy, nor is this some straightforward binary of sound and silence, or of voice overcoming the attenuation of state oppression. In his (banned) 2004 novel about life in Syria under dictatorship, *The Silence and the Roar*, Nihad Sirees depicts noise as a tool of tyranny and coercion:

Barely eight thirty in the morning and the sounds outside were all chaos. Sounds turned into noise as a bullhorn amplified a goddamned voice reciting inspirational poetry, utter gibberish that was only interrupted by the occasional barked instruction. The meaning of all those words got lost because another loudspeaker was simultaneously blaring motivational anthems. Meanwhile schoolchildren parroted the refrain, "Long live . . . Long live . . ."[52]

The cumulative effect of this sound clash is that "the roar produced by the chants and the megaphones eliminates thought."[53] Regimes endure through sonic techniques. In different ways, sound and music were also tools of dictatorship in Greece during the military junta in the 1960s and 1970s.[54] And in both places, as is the case elsewhere, the "soundwork" of uprising involves reclaiming audiopolitical tools and turning sound technologies toward new political futures.[55]

In 2011, women in Damascus installed loudspeakers on the roofs of buildings and in parks in the city, broadcasting revolutionary songs when the regime tried to block demonstrations.[56] Songs mocked the regime, and the regime found this intolerable.[57] Protestors turned romantic songs into antiregime anthems.[58] Ibrahim Qashoush/Abdul Rahman and Abdul Baset al-Sarout electrified movements in Hama and Homs with their vocal performances, creating a contestatory space that also became a space of circulation.[59] In the same moment, Athenian publics wedged loudspeakers into the branches of trees in the square below the Greek parliament building, as part of a summer-long occupation that remade the city and the way its political communities exercised vocality.[60] These practices resonate around Mediterranean cities and converge now in Athens.

Arada moves around the city. It is performed in streets and squares, in community centers, in front of parliament. It takes place in the same spaces where Greek publics have protested against austerity, against economic and military intervention, adding to the archive of resistance written into the city's urban fabric. And although this archive is fragile, ephemeral, something that needs continual maintenance, utterance, and performance, it continues to ring out across transnational and transurban geographies and has now outlived the regime it sounds out against. Revolution songs are being sung all over Syria again, these shared vocalities continuing to protect the cities and their people. The revolution remains a revolution— one that was scattered, its layers and vocalities accumulating and remapping across time and space—and has now, somehow, found its way home.

Gamal Abdel Nasser's Disembodied Voice

Anticolonialism and third worldism were sonically charged historical moments that brought together different geographies through a shared commitment to a decolonized future. Through revolutionary protest songs and chants, the sounds of armed struggle and liberation war, or the radio broadcasts spreading information about movements near and far, all of this and more meant that the sounds of anticolonial resistance were central to animating and creating new spaces of politics. These sounds did the political work of connecting spaces, people, and ideas that were tied together materially through experiences of colonialism and capitalism and ideologically through third worldism and anticolonial liberation; as noted earlier, we refer to this as *sonic third worldism*.

Radio broadcasts in particular were a crucial element of anticolonial struggle and connectivity.[61] Voice of the Arabs, for instance, was a foundational Egyptian radio station in the 1950s and 1960s that broadcast news, political commentary, speeches, music, and cultural programs about anticolonial struggles around the world.[62] The first episode aired on July 4, 1953, and the station is said to have been the brainchild of Egypt's postcolonial leader Gamal Abdel Nasser, who saw the radio as a pivotal tool in the struggle for liberation. By the 1960s it was broadcasting twenty-four hours a day, with an extremely large audience across the Arab world. On the other hand, there was Radio Cairo, which reached many countries across Africa. As James R. Brennan notes, "Broadcasts from Cairo offered a powerful vision of an emerging Afro-Asian world."[63] Radio Cairo was able to attract broadcasters and technicians away from colonial radio stations through their ability to pay employees higher salaries and was listened to widely, especially in East Africa, where transmissions were easily accessible.

The story of Voice of the Arabs and Radio Cairo also tells the story of the rise of Nasser, a core member of the Free Officers group that had spearheaded the revolution against British rule, leading to Egypt's independence in 1952. He presided over a popular project of decolonization, emphasizing political, economic, and social independence and underpinned by third worldism, Pan-Arabism and Pan-Africanism as a global project of transformation. Nasser became one of the most important postcolonial figures within third worldism, initiating dramatic changes within Egypt and across North Africa and the Middle East. Egypt's economy turned away from foreign capital and privatization, toward nationalization. Education and health care were made accessible to all, and investment into the public sector was prioritized. Nasser expressed both ideological and material support for anticolonial movements and postcolonial states across Africa and the Middle East, seeing Pan-Arabism and Pan-Africanism as central to Egyptian foreign policy. The 1950s and early 1960s were thus a monumental time in Egypt, during which dramatic political, economic, and social transformations took place.

The radio was important during these decades because some of Nasser's most popular speeches, including when he nationalized the Suez

Canal in 1956, were broadcast on stations such as Voice of the Arabs. Political speeches by anticolonial figures were central to the soundscape of anticolonialism, not least in Egypt, where Nasser's voice was known to almost every Egyptian and Arab, but also in many in other parts of the world. Nasser's speeches were extremely popular and affective, acting as rallying calls in the same way revolutionary chants did during the 2011 revolution. In the rest of this section, we explore the soundscape of one particularly powerful speech through its representation in a film, highlighting how the sounds of anticolonialism were constitutive of the political and social worlds of decolonization.

Essa Grayeb's *The Return of Osiris* (2019) is a film made up of other films. It collects different clips from a host of Egyptian films that documented or enacted Nasser's famous resignation speech, which he gave to the Egyptian public following the military defeat of the Arab armies by Israel in the Six-Day War in 1967. A central dimension of Nasser's political project was to liberate Palestine from Israeli settler colonialism, culminating in several military defeats for Arab states, including in 1967. The 1967 defeat was colossal, not least because Egyptian radio broadcasts had been publicizing Egyptian dominance during the short war. As such, the defeat came as a shock and has since been understood as a pivotal political moment in the Middle East, leading to the end of Arab nationalism and various anticolonial projects. Following this defeat, Nasser announced that he would resign.

Bringing together different clips from Egyptian films that included this famous resignation speech, Grayeb produces a complex sonic and visual experience layered with memory and history. Each clip brings its own soundscape and visual imagery with it, as well as a whole range of familiar Egyptian actors and actresses. We move quickly from one film to the next, with each film responsible for reproducing a small part of the speech. What ties this montage together is Nasser's voice, ringing through each scene except where silence or loud cheers take its place. Sometimes his voice is clear, other times muted; sometimes booming, other times quiet. Nevertheless it is always his voice that guides us as we try to make sense of what we are watching. The film is thus an example of how sound is a portal into anticolonial histories and how sonic experiences speak directly to the memories many Egyptians have of the anticolonial moment.

Nasser's resignation speech has been featured widely in Egyptian cinema and literature, given the ramifications of the defeat. Grayeb's montage is therefore both a film in and of itself and an archive of how the defeat was understood and represented when it happened and ever since. Grayeb, a Palestinian filmmaker and photographer based in Jerusalem, is currently working on a project about Nasser, of which this film is a part. Grayeb initially became interested in Nasser because of his childhood memory of seeing Nasser in Palestinian homes, including his grandmother's: "I remember asking people about the guy in the portrait and they were saying this is Nasser, this is Gamal. I lived my childhood thinking he's my uncle. I grew up with this memory, and then I found out Gamal Abdel Nasser is an uncle for many people. And I thought, what is this image of an Egyptian leader doing in Palestinian houses?"[64] These images of Nasser inside the home spoke to another fascination he had, of seeing these images in Egyptian films. He recounts seeing the same image of Nasser hanging on his grandmother's wall in a film starring famous Egyptian actress Suad Hosny. He became interested in using films about Nasser to explore questions of memory. *The Return of Osiris*, released in 2019, used extracted footage from Egyptian films and television series produced between 1976 and 2019 that include a part or all of Nasser's speech. Cumulatively the film, almost fourteen minutes in length, reproduces the whole speech. This is another anticolonial antiphony, a resonating voice.

In the film different people from across Egypt and across class, gender, and ethnicity gather around the radio to listen to the speech. The streets are completely empty, and there is no sound outside other than silence and the echoes of the speech from different radios. The scenes of silent streets, empty waterfronts, and deserted beaches are particularly powerful, juxtaposed against cities and towns that are usually loud and full of life. It is in the silence, even more so than in the speech and the reactions to it, that we might apprehend the gravity of the moment, moving through the speech seamlessly, each part coming just after the next, and yet we are watching vastly different fragments that have been tied together. There are differences in background noise, quality of the audio and video, music, and actors and actresses, yet somehow Nasser's voice and the speech are strong enough to act as a thread weaving the narrative together.

Because of the effect of Nasser's voice, the speech appears seamless despite the footage having been extracted from many different sources. The visual imagery of the film is strong, particularly given the familiarity of the faces and scenes taken from famous Egyptian films, yet it is Nasser's voice that moves us from one scene to the next. While there is a common refrain that most Egyptians know the speech by heart, Nasser's voice is even more embedded within memory given the popularity of his speeches and how often they are reproduced on screen and on the radio. Although he is not technically disembodied, and his image appears throughout the speech, there is a sensation of disembodiment that makes the film an important sonic archive of anticolonialism. As such, the film is a powerful archive partly because of the familiarity of Nasser's voice.

If memory can be understood as "an archived past that fulfils the need of preserving an affective relation with an exhausted past threatened with oblivion," then the sonic layers of the film speak directly to affect, temporality and memory.[65] The heightened affective state of both the moment of defeat and its recognition (via the speech) is captured precisely by the constant movement between sources in the film, never allowing you to quite settle down. The speech is heard across a whole range of sound qualities, from sharp to static, making the words seem close, far away, or somewhere in between. The experience of hearing the speech, as it plays with temporality, makes the past seem either far away or very nearby. Moreover, the viewer's memories of each film or television series are also entangled within the viewing experience, adding an additional layer.

The speech marked a moment during which Egyptians were confronted with the possibility of living without Nasser and life after anticolonialism. Nasser's announcement of his resignation suggested more than just life without him; it also meant a likely end to the political project of anticolonialism he represented. Indeed, what came after Nasser was a dramatic turn to neoliberalism and the controversial peace treaty between Egypt and Israel. As represented in Grayeb's film, the speech and the response to it constitute important sonic archives of anticolonialism that speak to the power of Nasser's voice, crowds demonstrating for him to stay, and the silence before and right after the speech. Because of how the film is put together, the sonic emerges most clearly, tying together disparate films and series. A pivotal turning point in Egyptian anticolonial history is thus experienced as a sonically charged series of events.

Sonic Third Worldism in Athens

> Εδώ Πολυτεχνείο! Εδώ Πολυτεχνείο! Ο αγώνας μας είναι κοινός. Είναι αγώνας αντιχουντικός. Είναι αγώνας αντιδικτατορικός. Είναι αγώνας αντιμπεριαλιστικός.
>
> (This is the Polytechnic! This is the Polytechnic! Our struggle is common. It is an antijunta struggle. It is an antidictatorship struggle. It is an antiimperialist struggle.)[66]

Greece, too, has histories of third worldism, also sonically charged and relayed in voices. The spectacular apogee of this came when students occupied Athens Polytechnic University as an act of collective resistance against the military dictatorship, known as the Polytechneio uprising of November 1973. The junta had been running the country since 1967—the same year as the defeat of the Arab armies by Israel. This occupation was a sound event. Over three long days, from November 14 to 17, the occupying students improvised a public broadcast system to communicate with the crowds that gathered outside the university, and they built a radio station that the whole of Athens could hear (and that the police and the military couldn't shut down).[67] In this last story, we hear the anticolonial politics of this moment and these broadcasts and earmark this as another type of sonic third worldism that resonated on global scales.

The Polytechneio uprising has been written about at length, a collective trauma written into the Greek psyche. It marks a violent climax of a dictatorship built on systematic repression, incarceration, torture, and murder of its political opponents. The student occupation was a culmination of growing dissent and resistance, which gathered into a mass mobilization in 1973, led largely by students and youth movements.[68] Its importance in Greece is hard to overstate. In Neni Panourgia's vivid account, "Out on the streets, inside the Polytechnic, at middle-class and working-class homes, this was the moment."[69] The junta lasted only until the following summer. Polytechneio marks a late act of the regime's brutality. In the early morning of Saturday, November 17, the generals sent tanks rolling through the city and sent one crashing through the university gates.

The radio announcements, relaying the student occupation to the city beyond, laid out the scope of the uprising as an antijunta struggle, an antidictatorship struggle, an anti-imperialist struggle. Rising steadily in intensity over the course of the occupation, the broadcasts called for all of Athens to come out onto the streets to bring down the junta. The imperialism in question pertains to the US support for the dictatorship, itself an extension of US interference in Greek politics since the World War II and its violent repression of leftist movements.[70] Out on the street, people did the work of connecting the different scales of resistance. On one level, chants were targeted directly at the needs of the political moment: "down with the junta"; "we want resistance"; "everyone united"; "fascism will not stand"; "bread, education, freedom."[71] On another level, the students in Athens placed this struggle into a wider anticolonial struggle playing out worldwide.

There were chants for ousted Chilean president Salvador Allende. Following the CIA-backed Pinochet coup in Chile two months earlier, students gathered in the center of Athens to shout Allende's name. The name itself became a slogan and was chanted again repeatedly during the Polytechneio uprising.[72] There were chants for Thailand. In October 1973, one month before Polytechneio, protestors in Bangkok occupied government buildings of the military junta—also backed by the United States. The crowds at Polytechneio responded with the chant, "Tonight there will be Thailand."[73]

These callouts, and the struggles and resistance movements they connect, are vocalizations of an internationalist cultural politics that precedes and exceeds these specific moments. Athens hosted the Anticolonial Conference of the Mediterranean in 1957 and became the base of the Permanent Committee for Anticolonial Struggle of the Mediterranean in the same year.[74] Moving against the grain of invocations of the city as some European wellspring, and complicating any clean geographical splits that divided the world into three during these postwar decades, another Athens sits within geographies of anticolonialism that circle the Mediterranean and echo from far beyond. For historian Kostis Kornetis, Greek youth movements in the 1960s and 1970s were part of radical cultures of the northern Mediterranean, in which third worldism was an inspiration and "revolutionary guide": "The struggle against Pinochet's dictatorship, alongside Vietnam and later on the anticolonial struggle in Portugal's colonies . . . , was placed on the same footing as the local struggle against the Colonels."[75] The radio broadcasts and chants at Polytechneio are another kind of sonic third worldism. By placing themselves into global anticolonial struggle, protestors in Athens made antiphonal spaces at multiple scales. The university radio and the streets outside were connected in a call and response; the chants to Chile and Thailand, and the invocations of Vietnam and Angola, bring Athens into global third-worldist movements; the squeal and squall of the radio transmission signal, searching for and making counterhistories, become powerful frequencies, plugging Polytechneio into far-flung relays of radio as a decolonial tool and technology.[76]

This moves us beyond thinking about Polytechneio as only a national struggle or a narrow contestation of US imperialism.[77] Polytechneio instead reveals a stereophonic political consciousness that connects struggles and communities through antiphonies that resound on a planetary scale, shuttling across sites and spaces in ways that confound colonial cartographies and neat descriptions of global North and South. Music, by this time, was being used as a tool to develop a cosmopolitan "third world consciousness" on a global scale—notably through the work of La Casa de las Américas in Cuba, which gathered, translated, and circulated protest songs from around the world.[78] And this builds on what Denning calls a "noise uprising" in the 1920s and 1930s, when recorded vernacular musics became a soundtrack to decolonization and sound figured anticolonial space as nonnational space.[79] Sound, for Denning, "makes possible new and unexpected reverberations, new forms of affiliation and solidarity across space and time."[80] Athens, and its conjoined port city, Piraeus, was part of an archipelago of interlinked places and spaces and of musical counterpoint that unmade imperial geographies of center and periphery—counterpoints that continued to weave through the twentieth century and that resound into the present.

Sonic third worldism shows how local struggles have long been connected through sound, looping with the crosscurrents that connect geographies and histories and hearing how sonic imaginations can trigger political change.[81] These are, in other words, audio equivalents of what Chandni Desai and Rafeef Ziadah call "insurgent geographies of connection," wherein struggles relate and respond to each on their own terms, making their own sense of space and movement.[82] In the recordings from the Polytechneio uprising, it is possible to hear and hold both the sense of defeat and the sense of enduring resistance. Gunfire rattles, the tanks crash through the university gates, the street falls silent, and the radio cuts out, leaving only the white noise of imperialism and the static of history. Yet the sound archive of Polytechneio continues to resonate through the city. The year the dictatorship fell, this archive expanded through the performance of music that had been banned or censored under the regime and through the voices of people shaping the politics of the city in the streets.[83] And still every year there is a march in Athens to commemorate this event, moving through the city and gathering outside the US Embassy. The chants provide the soundtrack that again ties past and present together and that join people in antiphonal resistance.

• • •

This article has explored antiphony and sonic third worldism as continuing anticolonial methods that reverberate into the present and the future. Moving between Cairo and Athens, we traced some of the promises and contradictions of anticolonial and revolutionary movements, picking up on the way how they speak to one another despite belonging to different places and eras. As such, sonic third worldism not only carries across time and space but also disrupts linear notions of temporality and geography in ways that attune us to connection and solidarity. These anticolonial antiphonies are crucial to highlight not only because they center sound and the emotions it can bring to the surface but also because they produce an archive of anticolonial connectivity that disrupts imaginings of Cairo and Athens as geographies that can be neatly separated. Instead, we have argued that sound offers a means through which to connect space—Cairo and Athens—and time: anticolonialism and the present.

Arada in Athens, a disembodied voice making its way through Egyptian streets, the chants and broadcasts that bind movements together —these anticolonial antiphonies remind us that resistance is ongoing and that hope is an important political practice. Sound attunes us to registers of anticolonialism that have otherwise slipped out of memory and history and yet continue to feed into the present. Sonic third worldism, though muffled and always under threat, continues to reach out toward decolonial futures—across the Mediterranean and as far as it can carry.

Sara Salem is associate professor in sociology at the London School of Economics. Her first book is *Anticolonial Afterlives in Egypt: The Politics of Hegemony* (2020), and she is currently working on a second book project on anticolonial archiving.

Tom Western's work builds creative geographies that seek to imagine futures beyond the colonial past and present. He works primarily in Athens, where he is involved in various forms of creative and collaborative research and movement building. He is currently finishing his first book, *Circular Movements*, and has recently published pieces in *Society and Space*, *Journal of Creative Geography*, and *Migration and Society*. He is based at University College London as a lecturer in social and cultural geography.

Notes

1. Gilroy, *Black Atlantic*, 79.

2. We are interested in building what Lisa Lowe calls "scenes of close connection in relation to a global geography that one more often conceives in terms of vast spatial distances" (*Intimacies of Four Continents*, 18).

3. In thinking the Mediterranean as a space of connection and contestation, we join with historians, geographers, and anti- and postcolonial writers who have explored the encounters and circulations that produce the sea. See Al-Mousawi, *Two-Edged Sea*; Chambers, *Mediterranean Crossings*; cooke, Göknar, and Parker, *Mediterranean Passages*; Hawthorne, *Contesting Race and Citizenship*; Khuri-Makdisi, *Eastern Mediterranean and the Making of Global Radicalism*; and Matvejevic, *Mediterranean*.

4. Here we follow Ana María Ochoa Gautier, who describes the work of "soundmapping" as building "an intellectual cartography that asks us to pay attention to the many possibilities of thinking in sound . . . —one that is less a map and more a conglomeration of stories" ("Afterword," 271).

5. El Alaoui, "Meaning of Bandung," 74.

6. Fanon, "This Is the Voice of Algeria," 94.

7. Fanon, "This Is the Voice of Algeria," 97.

8. Hill, *Black Soundscapes White Stages*, 141.

9. Denning, *Culture in the Age of Three Worlds*; Mahler, *From the Tricontinental to the Global South*; Phạm and Shilliam, *Meanings of Bandung*.

10. Because of the multiple forms of colonialism and imperialism involved in these stories, our use of the terms *colonialism* and *imperialism* throughout this piece are intended to contain European forms of empire stretching back to the fifteenth century, as well as the neocolonialism of US interventionism through the twentieth century and into the present—including its backing of forms of settler colonialism that are ongoing. As we articulate later, anticolonial antiphonies push against multiple forms of colonialism, and our use of terms thus reflects this transhistoricity and multipolarity of imperialism and struggles against them.

11. Desai and Ziadah, "Lotus and Its Afterlives"; Agathangelou, "Throwing Away the 'Heavenly Rule Book'"; Denning, *Noise Uprising*.

12. E.g., Alonso, "Broadcasting the '(Anti)Colonial Sublime'"; Hermosilla, "Gathering of the Protest Songs"; Karmy and Schmiedecke, "'Como se le habla a un hermano'"; yamomo, "Sonic Experiments of Postcolonial Democracy."

13. Ekomane, "Music of Anticolonial Resistance"; Hop, "De-Westernized Sonic Practices and Technologies"; Khazrik et al., "Fil Mishmish"; Shirhan, "Lovesong Revolution."

14. Khanna, *Visceral Logics of Decolonization.*

15. Colla, "Poetry of Revolt."

16. See Colla, "Poetry of Revolt"; LeVine, "Music and the Aura of Revolution"; Sanders and Visonà, "Soul of Tahrir"; and Ghanem, "2011 Egyptian Revolution Chants."

17. Colla, "Poetry of Revolt." This was an interesting broader trend, whereby the appropriation of old songs and chants was widely visible (Sanders and Visonà, "Soul of Tahrir").

18. LeVine, "Music and the Aura of Revolution," 794.

19. Sanders and Visonà, "Soul of Tahrir," 214.

20. Khanna, *Visceral Logics of Decolonization,* 23.

21. Khanna, *Visceral Logics of Decolonization,* 150, quoting Cesare Casarino.

22. Attalah, "The Sound of Prison."

23. Attalah, "The Sound of Prison."

24. Mossallam, "To Chant the Worlds Away."

25. Abu Hamdan, "Saydnaya."

26. Parker, "Forensic Listening," 149.

27. Parker, "Forensic Listening."

28. See Fanon's discussion in *The Wretched of the Earth* around how postcolonial elites often reproduce colonial dynamics rather than break from it (148–206).

29. Mossallam, "To Chant the Worlds Away."

30. Cola, "The Poetry of Revolt."

31. Sanders and Visonà, "Soul of Tahrir," 214.

32. Sanders and Visonà, "Soul of Tahrir," 214.

33. Mossallam, "To Chant the Worlds Away."

34. On "aural public spheres," see Ochoa Gautier, "Sonic Transculturation."

35. Halasa, Omareen, and Mahfoud, "Song in the Revolution"; Neggaz, "Syria's Arab Spring"; Issa, "Ibrahim Qashoush's Revolutionary Popular Songs."

36. Halasa, Omareen, and Mahfoud, "Song in the Revolution," 211.

37. Yassin-Kassab and Al-Shami, *Burning Country,* 58.

38. Yassin-Kassab and Al-Shami, *Burning Country,* 43.

39. Monzer al-Sallal, an organizer interviewed by Robin Yassin-Kassab and Leila Al-Shami, asserts that, through these chants, "we discovered the geography of Syria for the first time. If Homs was hit, in Manbij we'd chant 'Homs, we're with you until death.' It brought us all together. And now the values and traditions of different regions have been mixed together by migrating refugees. We've got to know each other better" (*Burning Country,* 170). As the authors themselves put it, "The most remarkable feature of the protest movement at this time was its ability to unite people across religions, sectarian and ethnic boundaries. The language of protest was neither religious nor secular; the demands as expressed on the street were for political rights to be applied in general, not to specific groups" (45).

40. Alkabbani, in Alkabbani and Western, "Movement Exists in Voice and Sound."

41. Yassin-Kassab and al-Shami, *Burning Country,* 36.

42. Khuri-Makdisi, *Eastern Mediterranean and the Making of Global Radicalism,* 144.

43. Al-Sawah, "Lure of the Street," 221.

44. These separations are often reproduced in academic study, which separates Middle Eastern studies and European studies from each other.

45. Theocharopoulou, *Builders, Housewives, and the Construction of Modern Athens,* 55–60.

46. Nakamura, "Untenable History."

47. Alkabbani and Western, "Movement Exists in Voice and Sound."

48. Syrian and Greek Youth Forum, "Al'Athinioun."

49. Palumbo-Liu, *Speaking out of Place.*

50. Joudah, "Topography of Gaza."

51. Harney and Moten, *Undercommons*, 118.

52. Sirees, *Silence and the Roar*, 8.

53. Sirees, *Silence and the Roar*, 20. The same point is made by Yassin-Kassab and al-Shami, who write that nationalist songs were an essential part of regime propaganda, played day and night in ministries and other public buildings (*Burning Country*, 40). And miriam cooke writes of the regime as a silencer, albeit through noise (*Dancing in Damascus*, 40–41).

54. Papaeti, "Folk Music and the Cultural Politics of the Military Junta in Greece"; Papaeti, "Popular Music and the Colonels."

55. On the audiopolitics of sound and revolution, see Denning, *Noise Uprising.*

56. Sahloul, "La Femme et la Revolution Syrienne"; cooke, *Dancing in Damascus*, 102.

57. cooke, *Dancing in Damascus*, 42.

58. Yassin-Kassab and al-Shami, *Burning Country*, 175. The work of artist Urok Shirhan speaks directly to this theme, particularly her piece "Lovesong Revolution."

59. cooke, *Dancing in Damascus*, 42–43, 64; Halasa, Omareen, and Mahfoud, "Song in the Revolution." The identity of the first singer hasn't been entirely clear. While the singer of revolutionary songs in Hama is widely known as Ibrahim Qashoush, who was murdered by the regime in 2011 as his voice was deemed too much of a threat, video has also circulated to suggest that the singer is actually Abdul Rahman Farhoud, now singing anti-Assad songs again following the fall of the regime at the end of 2024.

60. Stavrides, *Common Space*, 166–69.

61. See Thomson, "Worldmaking in the Palestinian Radio Stations"; Reza, "Reading the Radio-Magazine"; Potter et al., *Wireless World*; and Moorman, *Powerful Frequencies.*

62. Boyd, "Development of Egypt's Radio."

63. Brennan, "Radio Cairo and the Decolonization of East Africa," 174. This made them of particular interest to British colonial officers across Africa, as can be seen from the vast amount of material related to surveillance at the British National Archives.

64. Interview by the author, January 2022. Later, Grayeb adds that when he asked his cousins to describe their grandmother's living room (to confirm that the portrait he remembered did actually exist), they mentioned that next to the portrait of Nasser there had been a portrait of Grayeb's late uncle, which may be why he thought of Nasser as his uncle too.

65. Traverso, *Left-Wing Melancholia*, 84.

66. Papadakis and Bofiliakis, Εδώ Πολυτεχνείο.

67. Panourgia, *Dangerous Citizens*, 144.

68. Kornetis, *Children of the Dictatorship.*

69. Panourgia, *Dangerous Citizens*, 143.

70. The Greek National Army used US-supplied napalm to defeat the communists in the Greek Civil War of 1945–49.

71. Papadakis and Bofiliakis, Εδώ Πολυτεχνείο. This last chant, "Bread, education, freedom," is very close to the chant that gathered the demands of Egyptian protesters in 2011: "Bread, freedom, social justice."

72. Allende's name was also painted onto the university gates next to messages telling the United States and NATO to get out of the country. Kornetis, *Children of the Dictatorship*, 248.

73. Kornetis, *Children of the Dictatorship*, 248–49.

74. Kornetis, *Children of the Dictatorship*, 491; Stefanidis, *Stirring the Greek Nation*, 106–7.

75. Kornetis, "'Cuban Europe?'" 487, 510.

76. Fanon, "This Is the Voice of Algeria"; Moorman, *Powerful Frequencies*; Bronfman, *Isles of Noise*; Hill, *Black Soundscapes White Stages*.

77. Israel was also actively supporting the junta in Greece, including through economic and military cooperation, and the Greek dictators were reciprocally "happy about the glorious victory of the IDF" in 1967. Mack, "Suppressed History of Israel's Support for the Brutal Greek Junta."

78. Hermosilla, "Gathering of the Protest Songs."

79. Denning, *Noise Uprising*.

80. Denning, *Noise Uprising*, 233.

81. Denning, *Noise Uprising*; Shirhan, "Lovesong Revolution."

82. Desai and Ziadah, "Lotus and Its Afterlives," 297.

83. Papaeti, "*Songs of Fire* (1975)."

References

Abu Hamdan, Lawrence. "Saydnaya (the Missing 19dB)." 2017. http://lawrence abuhamdan.com/saydnaya.

Agathangelou, Anna. "Throwing Away the 'Heavenly Rule Book': The World Revolution in the Bandung Spirit and Poetic Solidarities." In Phạm and Shilliam, *Meanings of Bandung*, 101–12.

Alkabbani, Kareem, and Tom Western. "The Movement Exists in Voice and Sound." In *Sonic Urbanism: Crafting a Political Voice*, edited by Theatrum Mundi, 20–27. London: &beyond. 2020

Al-Mousawi, Nahrain. *The Two-Edged Sea: Heterotopias of Contemporary Mediterranean Migrant Literature*. Piscataway, NJ: Gorgias Press, 2021.

Alonso, Isabel Huacuja. "Broadcasting the '(Anti)Colonial Sublime': Radio SEAC, Congress Radio, and the Second World War in South Asia." *Modern Asian Studies* 57, no. 5 (2023): 1615–49.

Al-Sawah, Hani. "The Lure of the Street." In *Syria Speaks: Art and Culture from the Frontline*, edited by Malu Halasa, Zaher Omareen, and Nawara Mahfoud, 220–21. London: Saqi, 2014.

Attalah, Lina. "The Sound of Prison." *Mada*, April 2014. https://tinyurl.com/mx6zfpnr.

Boyd, Douglas A. "Development of Egypt's Radio: 'Voice of the Arabs' under Nasser." *Journalism Quarterly* 52, no. 4 (1975): 645–53.

Brennan, James R. "Radio Cairo and the Decolonization of East Africa, 1953–1964." In *Making a World after Empire: The Bandung Moment and Its Political Afterlives*, edited by Christopher J. Lee, 173–95. Athens: Ohio University Press, 2010.

Bronfman, Alejandra. *Isles of Noise: Sonic Media in the Caribbean*. Chapel Hill: University of North Carolina Press, 2016.

Chambers, Iain. *Mediterranean Crossings: The Politics of an Interrupted Modernity*. Durham, NC: Duke University Press, 2008.

Colla, Elliott. "The Poetry of Revolt." *Jadaliyya*, January 2011. https://www.jadaliyya.com/Details/23638.

cooke, miriam. *Dancing in Damascus: Creativity, Resilience, and the Syrian Revolution.* London: Routledge, 2016.

cooke, miriam, Erdağ Göknar, and Grant Parker. *Mediterranean Passages: Readings from Dido to Derrida.* Chapel Hill, NC: University of North Carolina Press, 2008.

Denning, Michael. *Culture in the Age of Three Worlds.* London: Verso, 2004.

Denning, Michael. *Noise Uprising: The Audiopolitics of a World Musical Revolution.* London: Verso, 2015.

Desai, Chandni, and Rafeef Ziadah. "Lotus and Its Afterlives: Memory, Pedagogy, and Anticolonial Solidarity." *Curriculum Inquiry* 52, no. 3 (2022): 289–301.

Ekomane, Jessica. "Music of Anticolonial Resistance." *Cashmere Radio*, July 2020. https://cashmereradio.com/episode/open-sources-27-music-of-anti-colonial-resistance-old-and-new-sonic-imaginations/.

El Alaoui, Khadija. "A Meaning of Bandung: An Afro-Asian Tune without Lyrics." In Phạm and Shilliam, *Meanings of Bandung,* 61–74.

Fanon, Frantz. "This Is the Voice of Algeria." In *A Dying Colonialism,* translated by Haakon Chevalier, 69–98. 1959; repr., New York: Grove Press, 1967.

Fanon, Frantz. *The Wretched of the Earth.* 1961; repr., London: Penguin, 2001.

Ghanem, Hiba. "The 2011 Egyptian Revolution Chants: A Romantic-Muʿtazilī Moral Order." *British Journal of Middle Eastern Studies* 45, no. 3 (2018): 430–42.

Gilroy, Paul. *The Black Atlantic: Modernity and Double Consciousness.* 1993; repr., London: Verso, 2022.

Halasa, Malu, Zaher Omareen, and Nawara Mahfoud. "Song in the Revolution." In *Syria Speaks: Art and Culture from the Frontline,* 210–21. London: Saqi, 2014.

Harney, Stefano, and Fred Moten. *The Undercommons: Fugitive Planning and Black Study.* Wivenhoe, UK: Minor Compositions, 2013.

Hawthorne, Camilla. *Contesting Race and Citizenship: Youth Politics in the Black Mediterranean.* Ithaca, NY: Cornell University Press, 2022.

Hermosilla, Matias. "The Gathering of the Protest Songs: Cuba, Third Worldism, and the Birth of the Protest Song Movement (1967–1970)." *The Global Sixties: An Interdisciplinary Journal* 15, nos. 1–2 (2022): 180–99.

Hill, Edwin C., Jr. *Black Soundscapes White Stages: The Meaning of Francophone Sound in the Black Atlantic.* Baltimore, MD: Johns Hopkins University Press, 2013.

Hop, Ale [Alejandra Cárdenas], moderator. "De-Westernized Sonic Practices and Technologies: Online Panel." YouTube, June 2021. https://www.youtube.com/watch?v=FFRFzXWRBLo.

Issa, Sadam. "Ibrahim Qashoush's Revolutionary Popular Songs: Resistance Music in the 2011 Syrian Revolution." *Popular Music and Society* 41, no. 3 (2018): 283–301.

Joudah, Nour. 2023. "Topography of Gaza: Contouring Indigenous Urbanism." *Jadaliyya,* May 2023. https://www.jadaliyya.com/Details/45043/Topography-of-Gaza-Contouring-Indigenous-Urbanism.

Karmy, Eileen, and Natália Ayo Schmiedecke. "'Como se le habla a un hermano': La solidaridad hacia Cuba y Vietnam en la Nueva Canción Chilena (1967–1973)." *Secuencia* 108 (2020): 1–33.

Khanna, Neetu. *The Visceral Logics of Decolonization.* Durham, NC: Duke University Press, 2020.

Khazrik, Jessica, et al. "Fil Mishmish ANTI-ANNEX ANTI-COLONIAL ANTI-RACIST Solidarity Mix." *Radio Alhara,* July 2020. https://soundcloud.com/jessikakhazrik/fil-mishmish-anti-annexation-anti-colonial-anti-racist-solidarity-mix-radio-alhara.

Khuri-Makdisi, Ilham. *The Eastern Mediterranean and the Making of Global Radicalism, 1860–1914*. Berkeley: University of California Press, 2013.

Kornetis, Kostis. *Children of the Dictatorship: Student Resistance, Cultural Politics and the 'Long 1960s' in Greece*. New York: Berghahn, 2015.

Kornetis, Kostis. "'Cuban Europe?' Greek and Iberian *tiersmondisme* in the 'Long 1960s.'" *Journal of Contemporary History* 50, no. 3 (2015): 486–515.

LeVine, Mark. "Music and the Aura of Revolution." *International Journal of Middle East Studies* 44, no. 4 (2012): 794–97.

Lowe, Lisa. *Intimacies of Four Continents*. Durham, NC: Duke University Press, 2015.

Mack, Eitay. "The Suppressed History of Israel's Support for the Brutal Greek Junta." *+972 Magazine*, April 2023. https://www.972mag.com/israel-support-greece-junta/.

Mahler, Anne Garland. *From the Tricontinental to the Global South: Race, Radicalism, and Transnational Solidarity*. Durham, NC: Duke University Press, 2018.

Matvejevic, Predrag. *Mediterranean: A Cultural Landscape*. Translated by Michael Henry Heim. Berkeley: University of California Press, 1999.

Moorman, Marissa J. *Powerful Frequencies: Radio, State Power, and the Cold War in Angola, 1931–2002*. Athens: Ohio University Press, 2019.

Mossallam, Alia. "To Chant the Worlds Away. The Anatomy of the 2011 Revolution." *Art of Assembly*, April 2021. https://art-of-assembly.net/2021/04/08/alia-mossallam-to-chant-the-worlds-away-the-anatomy-of-a-2011-revolution/.

Nakamura, Carolyn. "Untenable History." *Offshoot*, March 2022. https://offshoot journal.org/untenable-history/.

Neggaz, Nassima. "Syria's Arab Spring: Language Enrichment in the Midst of Revolution." *Language, Discourse, and Society* 2, no. 2 (2013): 11–31.

Ochoa Gautier, Ana María. "Afterword: Sonic Cartographies." In *Remapping Sound Studies*, edited by Gavin Steingo and Jim Sykes, 261–74. Durham, NC: Duke University Press, 2019.

Ochoa Gautier, Ana María. "Sonic Transculturation, Epistemologies of Purification, and the Aural Public Sphere in Latin America." *Social Identities: Journal for the Study of Race, Nation, and Culture* 12, no. 6 (2006): 803–25.

Palumbo-Liu, David. *Speaking out of Place: Getting Our Political Voices Back*. Chicago: Haymarket, 2021.

Panourgia, Neni. *Dangerous Citizens: The Greek Left and the Terror of the State*. New York: Fordham University Press, 2009.

Papadakis, Giorgos, and Manolis Bofiliakis. Εδώ Πολυτεχνείο: Ένα Ηχητικό Ντοκουμέντο (*The Polytechnic: An Audio Document*). Athens: Lyra, CD 3001, 1997.

Papaeti, Anna. "Folk Music and the Cultural Politics of the Military Junta in Greece (1967–1974)." *Mousikos Logos* 15 (2015): 50–62.

Papaeti, Anna. "Popular Music and the Colonels: Terror and Manipulation under the Military Dictatorship." In *Made in Greece: Studies in Popular Music*, edited by Dafni Tragaki, 139–51. London: Routledge, 2018.

Papaeti, Anna. "*The Songs of Fire* (1975): Sonic Narratives of Resistance and Collective Memory." *Twentieth-Century Music* 20, no. 1 (2023): 6–22.

Parker, James. "Forensic Listening in Lawrence Abu Hamdan's *Saydnaya (the Missing 19dB)*." *Index Journal* 2 (2020): 145–68.

Phạm, Quỳnh N., and Robbie Shilliam, eds. *Meanings of Bandung: Postcolonial Orders and Decolonial Visions*. London: Rowman and Littlefield, 2016.

Potter, Simon J., et al. *The Wireless World: Global Histories of International Radio Broadcasting*. Oxford: Oxford University Press, 2022.

Reza, Alexandra. "Reading the Radio-Magazine: Culture, Decolonization and the PAIGC's Rádio Libertação." *Interventions* 24, no. 6 (2022): 857–78.

Sahloul, Najwa. "La Femme et la Revolution Syrienne." *Souria Houria*, July 2014. https://souriahouria.com/la-femme-et-la-revolution-syrienne-par-najwa-sahloul/.

Sanders, Lewis, and Mark Visonà. "The Soul of Tahrir: Poetics of a Revolution." In *Translating Egypt's Revolution: The Language of Tahrir*, edited by Samia Mehrez, 213–48. Cairo: American University in Cairo Press, 2012.

Shirhan, Urok. "Lovesong Revolution." 2020. https://urokshirhan.work/Lovesong-Revolution.

Sirees, Nihad. *The Silence and the Roar*. Translated by Max Weiss. 2004; repr., London: Pushkin Press, 2013.

Stavrides, Stavros. *Common Space: The City as Commons*. London: Zed, 2016.

Stefanidis, Ioannis. *Stirring the Greek Nation: Political Culture, Irredentism, and Anti-Americanism in Post-war Greece, 1945–1967*. London: Routledge, 2007.

Syrian and Greek Youth Forum. "Al'Athinioun." *Citizen Sound Archive*, February 2020. https://citizensoundarchive.com/2020/02/27/alathinioun/.

Theocharopoulou, Ioanna. *Builders, Housewives, and the Construction of Modern Athens*. Athens: Onassis Foundation, 2017.

Thomson, Sorcha. "Worldmaking in the Palestinian Radio Stations (1965–1982): Revolutionary Love and Anticolonial Afterlives." *Global South* 15, no. 2 (2022): 99–116.

Traverso, Enzo. *Left-Wing Melancholia: Marxism, History, and Memory*. New York: Columbia University Press, 2016.

yamomo, meLê. "Sonic Experiments of Postcolonial Democracy: Listening to José Maceda's *Udlot-udlot* and *Ugnayan*." *Southeast of Now: Directions in Contemporary and Modern Art in Asia* 6, no. 2 (2022): 133–46.

Yassin-Kassab, Robin, and Leila Al-Shami. *Burning Country: Syrians in Revolution and War*. London: Pluto Press, 2018.

Radio Silence

*Black Technologies and the Sound of Britain's
Dying Colonialism*

Tao Leigh Goffe

The only good system is a sound system.
—Jamaican saying[1]

[Radio] is one of the means of escaping the inert, passive, and sterilizing
pressure of the "native" environment. It is, according to the settler's
expression, "the only way to still feel like a civilized man."
—Frantz Fanon, "This Is the Voice of Algeria"

Listen carefully enough and you can hear the sound of Britain's dying colonialism. Many of the global stations of the British Broadcasting Corporation (BBC) have been shuttered, or are in the process, due to austerity
measures.[2] As with any technology, the duality of radio means it has been
deployed both for the aims of state control and for radical liberation from
said oppressive forces. There is a correlation between the primacy and
decline of news radio and that of European imperial dominance. It is not
only that radio is a technology of a bygone era but also that it is so closely
attached to the colonial infrastructure through which it was built. There is
the duality of the civilizing mission and the decolonial mission amid evolving media ecologies from World War II to the present.

World War II transformed the tactical uses of radio as the global
speed of sound accelerated due to competition by the Allied and Axis
powers for geopolitical dominance.[3] Beyond the sonic brinksmanship, the
way those in colonized territories innovated sonic technologies, taking the
military machinery of sound into their own hands, is often overlooked in

Social Text 162 · Vol. 43 No. 1 · March 2025
DOI 10.1215/01642472-11573354 © 2025 Duke University Press

Western military and media histories. Sound may have dutifully carried the colonial propaganda of the civilizing mission, but the reception of these audio transmissions could not be reliably controlled.[4] By centering historical actors of color and how they received, interpreted, theorized, and produced sonic media, a new way of anticolonial listening becomes possible.

During World War II, Black Caribbean innovators tinkered with military sound technologies such as the public address system to create their own inventions, manipulating sound for entertainment and what became the technology of dancehall music. The sound system, an electronic array of loudspeakers used to play recorded music, is an original Jamaican invention, though this fact is not often acknowledged. Experience with operating and repairing radar (radio detection and ranging) and sonar (sound navigation and ranging) devices led to new senses of spatial knowledge and detection through soundwaves for Black veterans. Troops were mobilized based on how sound was used to detect enemy submarines and encoded enemy transmissions.

Military funding has always shaped research and development for technological advances as much as culture, from fashion to music. Beyond militaristic purposes, novel forms of entertainment and knowledge production emerged as new sonic communities and diasporas formed. New political subjects connected with one another through radio waves. With strategic knowledge and influence, imperial boundaries were becoming reorganized during World War II.

Uncredited Black sound engineers, radio broadcasters, and practitioners created sonic political communities attuned to the perverse realities of British colonial racism. The primacy of sound and Britain's dying colonialism can be traced through a series of actors and the way radio broadcasting has evolved from the post-WWII period to the post-Brexit present. Territories like those in the Caribbean that did not participate in active combat during World War II on their home soil were nevertheless enlisted and asked to prove their patriotism. They became activated in unexpected ways. The racial segregation of the armed forces, or color bar, unintentionally led to new circuits of Black knowledge production, power, and circulation by those remanded to desk duty, electrical engineering, and maintenance of war machinery. Arming Black people and other subjects of color with weapons in large numbers was a risk few colonial powers were willing to take during World War II, no matter how desperate the battle or how loyal the soldiers. Battalions of people of color (typically military volunteers) existed, but to a limited extent.

Trained as telegraph operators and electrical engineers, technicians of color subverted and jammed signals in Algeria, in Jamaica, and in Tahiti. What media theorists Sara Salem and Tom Western call "anticolonial antiphonies" articulates the vibrancy of the call-and-response between colonized nations echoing messages of sovereignty to one another.[5] Just as European empires contended with their colonies, the American empire deployed Native languages as a strategic advantage, recruiting Diné, Comanche, Choctaw, and other soldiers of Indigenous nations.[6] In these pivotal moments of deploying media and surveillance technology, empires were forced to confront the languages and peoples they had attempted to genocidally eradicate for centuries. The ongoing dynamic of Western colonialism calls for a reevaluation of how patriotism manifested during the war effort.

More than a century after the invention of the phonograph by Thomas Edison in 1877, the evolution of sound technologies traces the shifting geopolitical spheres of influence and coloniality. Charting radio's primacy in the 1920s and decline in the 2010s follows the rise and fall of British imperialism across a century.[7] Founded in 1922, the BBC was a commercial operation until it became public in 1927. The establishment of the BBC World Service in 1932 instantiated the British attempt to claim dominion over shortwave radio across its empire.

From his perspective as a broadcaster for BBC's Eastern Service, George Orwell recounts just how effective the civilizing mission of imperial radio was in World War II in the context of British India. In his 1943 essay "Poetry and the Microphone," he contemplated the role poets and the British intelligentsia had to play in the war effort.[8] As an Indian-born Englishman who had served in the Imperial Police in Burma in the 1920s, he had a perspective colored by the lens of colonial control. He wrote,

> Broadcasting is what it is, not because there is something inherently vulgar, silly and dishonest about the whole apparatus of microphone and transmitter, but because all the broadcasting that now happens all over the world is under the control of governments or great monopoly companies which are actively interested in maintaining the status quo and therefore in preventing the common man from becoming too intelligent.

Orwell saw hope in the instrument of radio despite government or commercial aims. Orwell defined the common man in a way that did not include the colonized peoples of the world. The wireless war expanded soft power sonically amplified media as a means of imperial control. The British sonic sphere of influence across occupied territories and independent nations through radio waves was designed to make compliant subjects of millions of people of color, preventing them too from becoming too intelligent.

How empire eavesdrops using audio infrastructure tools is inextricably linked to surveillance, imperial control, and espionage networks of intelligence. Though the British Empire may have established an electrified network of radio across its territories, sonic dominion could not be reliably controlled for propaganda in absentia. After World War II, the Cold War continued to be wireless, in a competition for the sphere of influence over colonized and formerly colonized nations. The first two decades of the twenty-first century have seen rapid planned colonial abandonment, neglect, derisking, and divestment from colonized nations because empires are simply too expensive to maintain.

Taken together, the writings of Afro-Caribbean thinkers and practitioners—Stuart Hall, Frantz Fanon, Una Marson, Hedley Jones—form a media theory rooted in the anticolonial context of their origins. The sovereignty of sound lies in its nature as an object that belongs to no one. Who can truly own airwaves? Colonialism exerts power through the illusion of ownership and the reality of dispossession, but sound cannot be held. Black diasporic subjects have long been using sound as a means to not only retain but reinvent African culture in the face of genocide. That genesis is theory—it is poiesis. It thrives through the feedback loop of a Black diaspora network of sound.

The sound of Europe's dying colonialism can be tracked through the evolving function of radio from World War II to the present. Colonial governance extends to sound with the totality that Peruvian sociologist Anibal Quijano defines as "the coloniality of power" in the present.[9] Sound networks had the power to permeate all aspects of life and to cement unexpected allegiances. The coloniality of sound also meant that decolonial possibility was embedded within the wiring. After World War II, the French colonial order faced this decolonizing, which Martinican intellectual Frantz Fanon articulated in his 1959 "This Is the Voice of Algeria," preceding Algerian independence in 1962.

Explaining how Algerians emerged as the protagonists in their narration of history through news media, Fanon theorizes across the African diaspora using the lens of psychology. He writes,

> The Algerian at this time had to bring his life up to the level of the Revolution. He had to enter the vast network of news; he had to find his way in a world in which things happened, in which events existed, in which forces were active. Through the experience of a war waged by his own people, the Algerian came in contact with an active community. The Algerian found himself having to oppose the enemy news with his own news.[10]

The French did not anticipate this sonic political activation through news radio. Showing the subversive potential of anticolonial radio networks, Fanon heard the power of the limitless horizons for homegrown Algerian news. Like Orwell, he observed how colonial governments utilized radio waves, but unlike Orwell, he also heard the decolonial potential. Sonic cultures and politicized communities would inevitably become part of the downfall of French colonial control.[11] The coloniality of sound immersed the colonial subject in a "vast network of news," as Fanon notes. The infrastructural channels designed as spy networks lay the groundwork but had the potential to be subverted as listening publics became more aware of deciphering the news and their role within it.[12] Such was a parallel dynamic for Britain and its territories after World War II.

While military histories are often narrated through written records, memos, battle plans, maps, and visual sources, taking a sonic infrastructural approach to imperial archives reveals how sound activated new political subjects with competing loyalties. The following examples trace the sound of Britain's "dying colonialism" and the eventual loss of the metropole's signal and its sonic sphere of influence through the BBC. From various sources, including the blueprints of BBC's Bush House and 1950s radio technology magazines, I read how the colonial message became distorted by Black Caribbean technologists who manipulated sound toward their own decolonizing ends. Listening, instead of looking, to the decline of British Empire through the analytic of media technologies addresses contemporary debates on the limits and promises of state-funded media. The failure to adapt to the new media ecology of journalism has meant the failure to thrive for British radio. However, I would argue that radio, in general, is constantly reinventing itself in sonic revenant forms that appear in today's media landscape of social media, television, and podcasts.

Hedley Jones: Systems of Black Sound

The Jamaican saying, "The only good system is a sound system," reflects something of a Rastafarian ethos of rejecting colonial structures of governance and isms, and Babylon, or the West, the heart of corruption. What sound carries is integral to systems of Black sound in the Caribbean. Hedley Jones was a progenitor among Black inventors and audio pioneers who contributed to the mechanics of what would become the sound of Rastafari in the 1960s and 1970s: reggae. Jones, who served in the Royal Air Force as a volunteer airman (radar engineer), invented the sound system in the late 1940s in Jamaica. He had specialized trained in wiring and circuitry at the Glasgow and West of Scotland Technical College (Royal College of Science and Technology) during the war. He was a *boffin*, the British vernacular term for an engineer whose skills came into play during the war effort in unexpected ways. The term emerged to describe British Royal Air Force technicians who helped develop radar technologies. They were often the codebreakers on whose expertise the war effort depended, and Black servicepeople were not imagined to be among them.

Jones flourished in ways that perhaps did not especially seem valuable to the British military but would advance the Jamaican entertainment industry. He went on to adapt the technology of the public address system for the purpose of playing amplified recorded music for partygoers. By finding the right equilibrium, vocals could be emphasized. Alternatively, the pulsation of the bass could be emphasized by adjusting knobs. Jones's specific invention was engineering a way to manipulate levels—the high, mid, and low frequencies—of the amplification device for sonic enjoyment.[13] Prior to this, big band, jazz, and orchestral live music predominated as popular musical entertainment. With the amplified technology of the sound system came new methods of sonic manipulation for broad and often unintended audiences, especially as equipment became more affordable.

Jones described his speakers as "houses of joy." The architecture Jones created became a "structure of feeling" for Caribbean independence from Britain—Raymond Williams's term helps describe the affective way the sound system facilitated a spatial unboundedness for sound to carry beyond confines set by colonialism.[14] Black consciousness and relationships were formed by the social subjects brought together by the sound system and its liberatory range across fragmented island geographies. By now the blueprint of his engineering has spread across the globe, and the legacy is the proliferation of electronic dance music, as musicologist Michael Veal points out.

Returning to his home island from the United Kingdom after World War II, in 1949 Jones engineered the first speaker design to optimize levels for playing recorded music. A Kingston party promoter, hardware store owner, and deejay named Tom Wong, or Tom the Sebastian, asked Jones to engineer it so the lyrics could be heard, but also to emphasize the bassline. Wong, who was of Afro-Chinese heritage, became a conduit of access to bourgeois Chinese capital to adapt a Black sound and sonic practice that would spread to the Jamaican masses. While Jones was not a wealthy man, he had access to hardware and worked as a radio repairman, as well as doing general odd jobs and handywork. The skill of cabinetmaking was essential to the development of the wooden architecture of the sound system. He opened a record store, Bop City, in 1947, and Wong was one of his first customers. Out of this new sonic culture of playing recorded music, in which R&B songs from the United States and Latin music from the Spanish Caribbean were popular, reggae formed as a new homegrown Jamaican genre. It became the soundtrack of anticolonial defiance and Black power that the British government could not have anticipated would grow alongside formal Jamaican independence.[15]

Without the electrical innovation of Jones and promotion by Wong, the sound could not have traveled as fast as it did through urban and rural parts of the Caribbean. Emblematic of a Black technology, the sound system is what African American poet Imamu Amiri Baraka wrote of in his 1960 essay "Technology and Ethos." On the need for Black systems of sound, Baraka asks, "How do you communicate with the great masses of Black people?" He states, "Black creation-creation powered by the Black ethos brings very special results."[16] Under conditions of austerity, under British and US imperial influence and neglect, the Caribbean nevertheless became a hub of sonic innovation because of the ethos of Black liberation. Certainly, the sound system invented by Jones in Jamaica is an example of just such a Black creation, a Black technology amplifying the range and possibility of Black life, leisure, and pleasure.

Jones's sound system design was also influenced by his craft as a musician, a bass guitar player, leading the Hedley Jones Quartet. There are many archival artifacts that visually document Jones's process throughout his long career. Continuing into the 1970s, he was featured in local Canadian media for his sonic innovations with music instruments and the latest in sound technology and experimentation (fig. 1). A lifelong inventor, Jones went on to create the world's first solid-bodied electric guitar, among numerous other patented inventions, using timber endemic to Jamaica. Though he never published it, Jones sought to write his own life narrative and was frustrated by musicologists and their reluctance to engage with the science and especially the physics of sound. Without an understanding of basic mechanical engineering of how deejaying works, it is difficult to comprehend just how genius and singular Jones's inventions were. He trained several apprentices (e.g., Duke Lawrence, Oval Lue, Neville Cha Fung, Fred Stanford, Ucal Gillespie, Jackie Eastwood), so his expertise traveled as his students set up their own Black and Asian diasporic networks of sound. Some worked in entertainment, while others went on to work in electronic design at NASA and in media communications. Lawrence became chief engineer for the British Overseas Airways Corporation in London. Another audio engineer, Arthur Hassan, became an operator for the Jamaica Broadcast Corporation. Others he trained traveled to Brooklyn, bringing the technology of the sound system with them across Black diasporic circuits.

In addition to his service in the Royal Air Force, Jones learned much from the circulation of printed culture on radio technologies. He pored through magazine subscriptions of the period dedicated to radios, broadcasting, and mechanical and electrical engineering. Jones had taught himself the basics of radio and transistor tinkering before the war and was able to continue this education in Jamaica as an avid reader. Though not the intended audience for *Wireless World*, Jones learned from articles with tips on how to manage sources of interference. He applied and adapted these methods to the mountainous terrain of Jamaica, which had histori-

Figure 1. Hedley Jones Quartet using musical instruments and sound recording equipment.

cally made the transmission of certain radio frequencies difficult on the island. Topics of articles in *Wireless World* ranged from discussions of transistor symbols to the cloak of security, showing how entwined the sphere of the hobbyist and state surveillance were during the Cold War. When he could afford to, Jones would mail-order tools and supplies from these magazines. In his writing, he describes how his socioeconomic status prevented him from access to the latest technology in Jamaica. There were wealthy Jamaicans, he noted, who had connections outside the island and were able to get the newest sound equipment shipped directly from Miami. Nevertheless, the conditions of austerity for average Jamaicans became the conditions of possibility for new sonic inventions that would benefit the world decades later.

Eventually, reggae arrived in the metropole. Though stereo sound was becoming popularized in the 1960s, mono sound was the preferred sound system aesthetic based on the technology. Using one channel of sound, it is a distinctive feature of the speaker systems adapted to venues such as local house parties, basement clubs, and later concert halls. Depending on the type of track, the instrumentation of mono can sound crowded to some. But in reggae and dub music performance, the visceral experience of sound, and the feeling of the bass, means being open to other frequencies and ways of listening that involve a whole-body experience. Embracing mono reflects again the inventiveness of Caribbean migrants making the system and its mechanics their own via tinkering, engineering, and performing.

As Veal writes, the Caribbean was considered to be technologically lagging behind after World War II.[17] Yet many forms of sound technology innovation developed during the postwar period, especially in the Caribbean, despite conditions of austerity. It was Britain that was lagging behind musically. Sound was shattered and reinvented in Jamaica, as Veal argues. He goes on to explain that part of the signature of dub music that evolved out of reggae is an almost "unfinished" sonic quality. While at first this was criticized by non-Jamaican listeners, it soon became the defining aesthetic of the genre. Distorting, modulating, and elongating sound, reverb was the effect, stretching time, requiring an experimental listening experience. The conditions of possibility existed in the economic abandonment and neglect by the British Empire in places like Jamaica. The sonic invention of new genres and politics was possible because of Jones's blueprint, both as a method and as a machine of manipulation.

As 1950s music transitioned to sound systems as well as discotheques for popular entertainment, sound evolved with space from jazz to rhythm and blues to pop. With the Windrush era and migration of Caribbean people en masse to the United Kingdom, a great number of whom were band musicians, the music traveled and evolved, too. The spatiotemporal poetics of the advancing technology had a distinctly Caribbean DIY innovative nature expressed in the pulsing bassline of London basements. Regionally, local deejays and emcees in Bristol, Birmingham, Huddersfield, and Leeds became the diaspora transistors themselves, channeling the power of the sound system for partygoers. This dynamic was active across Britain and reverberated from Jamaica across the Caribbean to the United States and on to the United Kingdom in a diasporic triangulation of sound.[18] It was a Black diaspora network that formed in parties and in politics from an ethos of Blackness fragmented. While in many ways that soundtrack has been co-opted, it still reflects the origins of a radical politics of liberation from corrupt systems of imperial control.

Black diasporic consciousness emerged during this period in ways that Stuart Hall has poignantly articulated. Born in 1937, he was of the generation that migrated to the United Kingdom after World War II, as Jones, born in 1917, was returning to Jamaica from service in the United Kingdom.[19] The way Hall describes the emergence of Blackness in these fraught political moments transcends the simple epidermal fact of color as new politics of race emerge. Hall's landmark essay on modern media systems, "Encoding and Decoding," theorizes the matter of interference and reception in ways that Jones would have understood intuitively as a tinkerer. Emphasizing the labor process of media production and reception makes transparent the politics, circulation, and distribution of media. Hall writes, "Thus—to borrow Marx's terms—circulation and reception are, indeed, 'moments' of the production process in television and are reincorporated, via a number of skewed and structured 'feedbacks,' into the production process itself."[20] These words allow us to grapple with the full extent of Jones's labor as a British colonial subject and veteran becoming Jamaican.

Una Marson: Encrypting/Decrypting the *Voice of the Caribbean*

Centering Black labor in media spaces, there was no more impactful example of a BBC radio program than *Caribbean Voices*. Airing from the early 1940s to 1958, and broadcast from Bush House in London, the weekly show began with soldiers reading messages to their families. When Afro-Jamaican broadcaster Una Marson became the producer, the program developed to reflect her interest in poetry. The famed Jamaican intellectual and journalist was the first Black producer on the BBC payroll, and her role is all the more notable as a woman. Marson had been working for the BBC in London from 1939.

Sound was an intimate part of the emergence of Blackness in Britain through sanctioned and unsanctioned means, as Hall describes. While decentralized networks of literary knowledge production in print culture are more apparent and studied, sonic pathways of media infrastructure in the 1940s and 1950s existed, too. These modes are evidence of the power of alternative literacies and audiences that became activated beyond "received pronunciation" of the Queen's English.

Caribbean Voices is emblematic of the unknowing amplification of transmissions with anticolonial potential by radical Caribbean thinkers such as Samuel Selvon, Sylvia Wynter, Andrew Salkey, and Kamau Brathwaite.[21] Marson presents a patriotic newsreel in 1943 called *Hello! West Indies / West Indies Calling.*[22] The British voice-over emphasized that Caribbean people are of more than a dozen different races, depicting loyal white, Black, Indian, and Chinese people from the Caribbean in London as supporters of the Allied cause, happy to be in uniform. The newsreel features a military band of Black jazz musicians from the Caribbean and transitions into a montage of scenes of interracial cooperation between white British soldiers and Black technicians.

The manual labor of Caribbean people is emphasized by showcasing the Afro-Trinidadian cricket star Learie Constantine, who describes factory work and expertise being taught side by side with English men and women. The machinist training of the domestic production line transformed Caribbean workers into parts of the military-industrial complex. Constantine served in the war effort as a welfare officer in the Ministry of Labour and National Service. He later became a lawyer and a politician and was used by the Crown to make a future role within the British trade union system seem promising for Caribbean people hoping to migrate to the Mother Country.

In the newsreel, a calypso ditty with the refrain of "march to victory" reverberates, written to encourage recruitment into the army and Royal Air Force. The potential double meaning of calypso as a satirical genre often used for derisive or biting sociopolitical commentary would have been lost on white Britons. Caribbean technicians wear headphones and control electronic equipment as planes are being deployed to "fight Hitler and his gang," as the voice-over states. Here we see how the technical skills gained in engineering, soldering, and drilling led to a generation of trained West Indian artisans with experience in physics, mechanics, and fabrication, building on prior skills. The British state and war effort sanctioned this safe view of patriotic colonial subjects.

Another key figure in building Caribbean literary infrastructure during this period, Bajan poet Kamau Brathwaite said that *Caribbean Voices* was the "the single most important literary catalyst for Caribbean creative and critical writing in English."[23] The defiant seedlings of the Caribbean arts movement germinated through sound, cast broadly, in this moment across islands that had never heard one another before. Brathwaite noted the aurality of the radio experience and the feeling of interisland and diasporic connection facilitated by the radio technology in real time for British colonial subjects during World War II. It is noteworthy that he mentions not just literature but literary criticism and interpretation in *Caribbean Voices*. Brathwaite explains that what was cultivated on colonial radio waves of the BBC was a culture of discourse in messages sent to and from the Caribbean archipelago in what were on the surface simply patriotic broadcasts.

Brathwaite's emphasis is precisely on the sonic embodiment of Caribbean accents heard on British radio waves as having anticolonial potential, broadcasting across Britain. Marson played a large role in producing these segments and orchestrating the sound. Furthermore, for those who were not traditionally literate, the medium offered alternative pathways to education through listening and repeating. The affirmation of hearing Caribbean accents transmitting and being received worldwide sent a powerful message of recognition across the empire. The demand and desire for Caribbean labor by the British was also loud and clear, though the future reality of Black British life would be one of disavowal and violence.

Documentation of Marson's experience from a confidential 1941 report shows the bias against people of color by the white British authorities. While her abilities were praised to an extent, her technical skills and attitude were harshly critiqued.

> She is weak, however, on the mechanical side, and we imagine that as a producer she may be a shade impatient. At the moment, however, her main difficulties probably arise from social rather than technical circumstances. As the only "coloured" producer in the BBC, she needs a good deal of backing to enable her to use her abilities to the best advantage, and to overcome the prejudices which undoubtedly exist among some of the staff, with whom she has to work.[24]

Negotiating the scripts of race, nationality, diaspora, and competing allegiances, the BBC forms a backdrop for exploring generational shifts and attitudes about race and specifically Blackness. Marson's technological aptitude and capacity are questioned as much because she is Black and because she is a woman. She was no stranger to battling sexism, and as a Black feminist Marson was vocal globally about women's rights. She had spoken as the Jamaican representative at the International Alliance of Women for Suffrage and Equal Citizenship in Istanbul in 1935. At home in Jamaica, she was an ardent advocate of women's reproductive rights.

The soundtrack of London during the 1940s during the Battle of Britain was one of constant bombing that left the metropole in ruins, and Marson heard it loud and clear. After the war, Caribbean people were beckoned to fix the city in rubble. The British Nationality Act of 1948 was passed to recruit labor from across the empire to repair London. Poet Louise Bennett wittily referred to Caribbean people in the late 1940s Britain as "colonizing in reverse."[25] In her poem, Bennett's wordplay visualizes the confrontation between subjects and citizens in the British metropole. The narrative foreshadows the resentments of immigrants that shaped the anti-immigrant rhetoric of Brexit in 2016.

British Broadcasting's Blueprints

Fast-forwarding to the contemporary context, in 2011 the BBC shuttered many of its radio stations, including the Caribbean Service and the Portuguese for Africa Service. At the end of the final broadcast from BBC's Bush House, on July 12, 2012, the imperial lease on Caribbean radio waves and many other global stations, including former Soviet satellite nations, expired too. Then Prime Minister David Cameron's British austerity measures led to the mass defunding of large parts of the BBC World Service. A precursor to Brexit, his actions signaled what would be the future disinvestment and derisking of the Caribbean. However, Caribbean subjects have always responded to austerity with inventiveness, something the metropole will never be capable of.

The year 2011 was a pivotal moment of global unrest that included violent eruptions in the United Kingdom following the uprisings of the so-called Arab Spring.[26] The August 2011 uprisings in Tottenham were part of the sociopolitical climate of Cameron's attempts to harness state control. Leading Britain to war with Libya in the same year, Cameron demonstrated the Conservative coalition government's commitment to ongoing imperialism and Anglo-American friendship.

In this context, let us take Bush House, the former headquarters of BBC, as the locus of a colonial sonic project of broadcasting. Located near Holborn between the City and West End, that Cameron would choose to cut Britain's losses by downsizing here is full of symbolism. Radio has always been an intimate part of that colonial linguistic project.[27] Etched in the stone of the building are the words "To the friendship of English-Speaking People." A statue by American sculptor Malvina Hoffman symbolizing Anglo-American friendship looms.[28] Built in 1929, Bush House represents a sentiment that continues to echo of the friendship and special relationship between the US and United Kingdom that undergirds Anglophone dominance. The building does not bear the BBC motto, "Nation shall speak peace unto nation." Named after wealthy US mogul Irving T. Bush, the building is a testament to US-British collaboration. He funded the construction of the grand building, hoping it would be a major trade center. It was declared the most expensive building in the world.

The architecture of sound in the blueprints of Bush House takes the shape of surveillance and secret intelligence. There has long been an overlap between espionage and journalism as a convenient cover, and Bush House was romanticized as the United Nations of broadcasting.[29] It has also been dubbed a "benign Tower of Babel."[30] Anglophone global dominance has been a successful British mission for standardization of English as the language of business and diplomacy. The resonance and intonation of received pronunciation, its engrained racial and class politics, and the King's English continue to be part of the auditory tactic encoded in the radio and television broadcasts of the BBC.

The edifice stands for the notion that those in non-Western countries would need to turn on their radios to understand what was happening in their countries. Meanwhile, much of the political instability of the Cold War was the engineered interference of misinformation from the West. The ongoing colonial cycle of impoverishment and extraction was bolstered by sonic dependence, reinforced by the wiring and infrastructure of British media technologies. Local reporters were recruited to be the mouthpieces of the colonial message under the guise of objective journalism. At times individuals would be able to express the nuance of regional politics; however, editorial and production decisions were most often determined by British station chiefs trained in the metropole with a veiled agenda.

While the BBC continues its international reporting mission, the Tory-led decision about where geographically to divest first speaks volumes. Now that the Tory-led government has been disrupted, what happens next with regard to national radio remains to be seen. Just as we are currently witnessing financial derisking by the imperial Canadian banking institutional infrastructure, which is physically exiting the Caribbean for larger markets in South America, for the Tories the same region is the first to be abandoned.[31] In the imperial cost-benefit analysis, the Caribbean will always be a "small place" demographically. Thus, financially its markets are not seen as valuable, as was the case for the Cameron government. These actions, a denial of empire, were the bellwether for the sentiment that led to the Brexit referendum. The disavowal of the Caribbean and the wealth violently extracted by empire can be conveniently forgotten by turning off the radio. This is the reliable amnesia of colonialism.

While in the United Kingdom the BBC took the central role of media news and information services, telecommunications in the United States were always predominantly a private enterprise, with CBS and NBC the two main competitors.[32] During World War II, both channels became government vehicles for transmitting to the nation and empire. Contemporary debates about "state-affiliated media" belie the definition of propaganda. Funded by taxpayer monies, national media continues to have contentious roles across many countries. In both the United States and United Kingdom, broadcast media, especially radio, has become a primary target

Figure 2. BBC Heritage Trail plaque commemorating the Bush House.

for budget cuts. The rationale for conservatives to divest from government spending on media in favor of private corporations is consistent.

Prime real estate, Bush House was purchased by a Japanese company in 1989 and acquired via a fifty-year lease by King's College London as part of the expanded Strand campus in 2015. Repurposed by an architecture firm that claims to have maintained the integrity of the art deco edifice, the walls have so many layers of imperial desire and decline embedded in them. When the BBC's lease ended before it was leased to King's College, corporate owner Kato Kagaku Ltd. rebranded it as Aldwych Quarter.[33] While the inscription and original neoclassical design are intact, there was no attempt to retain the radio or media history of the building. There are no obvious markers of the BBC World Service except a simple blue plaque that is part of the BBC Heritage Trail (fig. 2). It is now a chapter of London's World War II history paved over and left behind.

The architectural palimpsest of governance through radio technology has transformed into part of the British colonial education-industrial complex, which depends on international student tuition revenue. Both institutions—universities and radio—are essential to the colonizing mission as civilizing mission, working hand in hand. As mass strikes and collective action are organized across UK universities amid rising tuition fees, salary stagnation, precarious academic contracts, and pension dismantlement, much is fraught in the structure of education that Britain could once rely on as an unquestioned standard-bearer.

Brain drain continues, disciplining those students deemed the brightest to become loyal and aspirational colonial subjects. The abandonment of radio infrastructure by Europe is part of the contemporary colonial reality of neglect for many parts of the world. While not the subject of this article, there is a long, flourishing history of unsanctioned soundwaves, namely, pirate radio, and other decentralized forms of media communication. There have been a number of public experiments, including the Jamaica Broadcast Corporation, a state-owned and statutory company that lasted from 1959 to 1997, as part of an effort founded by Jamaican Premier Norman Manley. Like the deferred dream of federation for the Caribbean, the dream of a pan-Caribbean media network holds promise for the fragmented archipelagic geography.

There was a hiatus at Bush House from the early 1950s to the late 1980s in dedicated radio programming focused on the Caribbean.[34] The BBC World Service's modern Caribbean Service started in July 1988, and its end on March 25, 2011, was abrupt. The shell of its website serves as a Cold War archive, with select new briefs and timelines of invasions, revolutions, and dictatorships.

Born in the United Kingdom to Jamaican parents, Colin Grant reflects on his experience as a BBC producer in *Bageye at the Wheel: A 1970s Childhood in Suburbia*, which features Bush House.[35] In his most recent book, *I'm Black So You Don't Have to Be: A Memoir in Eight Lives*, Grant addresses the civilizing mission at the heart of empire and what it meant to work for the BBC in light of the 2020 Black Lives Matter movement.[36] While the title of the memoir may sound postracial, it provokes a conversation about the meaning of race for the grandchildren of the Windrush generation. I heard Black British dub poet Linton Kwesi Johnson describe this generation as the grandchildren of Thatcherism, which is significant considering austerity and racial politics, too. Grant recounts that in the 1990s at Bush House he was unfairly accused of being aggressive in the workplace because of racial politics. Grant's contemporary account helps us read against the archival grain of how the BBC admonished Marson forty years prior.

Could colonial subjects listening after the victory of World War II have imagined that in London the BBC's lease at headquarters could ever be up? In the period just before it shuttered and was redesigned, Bush House was reportedly infamous for a mouse infestation could no longer be kept at bay. In disrepair, it was covered with sixty years of dirt layers as the doors closed—could the Windrush generation have imagined that the office furniture, down to the desks and chairs, would be auctioned off to the highest bidder? While the excuse of a merger or consolidation is the official alibi, the ruins in the metropole tell a story of insolvency. As it turned out, there had been nothing ceremonious about where the signal emanated from as it faded away.

By the end in 2012, BBC employees at Bush House described the carpets as worn out with holes and the marble floors as dented.[37] Dead air is perhaps the condition of possibility for a new homegrown sonic transmission in the Caribbean, Africa, and Asia. It used to be said that the sun would never set on the British Empire, spanning across time zones. What of the day of the last BBC transmission abroad? The radio silence of the colonial broadcast need not be lamented. A sovereign sound can only be born from new and decolonized infrastructures, invented through an ethos beyond the logic of colonial possession, as Una Marson and Hedley Jones show us.

What Sound Carries after British Empire

The ear cannot be decolonized without undoing the psychic damage of the toll of colonial propaganda, also known as colonial historiography. While Britain's colonialism is dying, the war is ongoing. Its holding of territories, many used for money laundering across the Caribbean, evince the range of economic control it refuses to relinquish. The twenty-first century requires a reevaluation of how we define both war and its media technologies. What role is there for radio after empire? What would it mean to truly create one's own news, as Fanon suggested, without it becoming another form of propaganda? Inventing new technologies, new media, and new philosophies, sonic cultural producers Marson and Hedley Jones set the blueprint.

Considering the long history of Caribbean occupation, from the sixteenth-century European invasion onward, the fragmented structure of the islands has always been a strategic military advantage. It has also been used to divide oppressed peoples. How could novel sonic technologies unite the nations of the archipelago and surrounding coastal countries? With homegrown broadcasts from the margins and the periphery from the post-WWII period, there is creative potential in what Veal calls the "shattering" and what I see as reinventing of that sound. The Caribbean has sustained itself through cycles of emergency. Because of its geography, it has been a dynamic sonic territory, picking up multiple unintended radio signals from South America, Central America, Britain, and the United States.

A twenty-first-century sonic cartography of interarchipelagic relation has potential. It exists in polyrhythmic scales of relation beyond the comprehension of European classical music. Sonic sovereignty could mean reorganizing political communities through sound. How can island territories tune into each other's frequencies without US or Western European interference? Out of the ruins, new media technologies continue to emerge from the unfinished project of Black reinvention that have never sought licenses or permission.

Tao Leigh Goffe is associate professor of literary theory and cultural history at Hunter College, City University of New York. Her first book is *Dark Laboratory: On Columbus, the Caribbean, and the Origins of the Climate Crisis* (2025). Her articles have been published in *South Atlantic Quarterly*, *small axe*, and *Women and Performance*.

Notes

Many thanks to editors Tariq Jazeel and Tom Western for inviting me to be part of this intellectual community on the interdisciplinary range of sound studies and geography. I'm thankful for the funding made possible for the opportunity to travel to the *Sound Carries* conference at University College London. Spending time with military archives and in the Imperial War Museum was invaluable to writing this article and transforming my view on the Royal Air Force in the Caribbean, as was time spent at the Library of Congress Recorded Sound Research Center.

1. I most recently saw this saying hand-painted on a sound system at Dub Club in Kingston, in 2019.

2. By 2030 the BBC plans to become an online-only service. Over the years it has faced many cuts to local and regional news outlets. In 2011, Albanian, Macedonian, Serbian, English for the Caribbean, and Portuguese for Africa shuttered.

3. After World War II, a flood of radio technology advancements blossomed as inventions disseminated beyond national boundaries. Great Britain and the United States greatly benefited from German innovations in magnetic tape recording.

4. Stuart Hall, "Encoding and Decoding in the Television Discourse."

5. See Salem and Western's contribution to this special issue, titled "Antiphonal Antiphonies."

6. See Connole, "Comanche Code Talkers of World War II."

7. Denning, *Noise Uprising*.

8. Orwell, "Poetry and the Microphone."

9. Quijano, "Coloniality of Power, Eurocentrism, and Latin America"; Mignolo, *Darker Side of Western Modernity*.

10. Fanon, "A Dying Colonialism," 4.

11. Fanon, "This Is the Voice of Algeria."

12. Satia, "War, Wireless, and Empire."

13. Goffe, "Bigger than the Sound."

14. Williams and Orrom, *Preface to Film*.

15. In this distinction between formal independence and sovereignty, I'm led by

the cogent articulations of Deborah Thomas, David Scott, and Donnette Francis from the convening of their Jamaica 1950s conference. See Thomas, "Displacements."

16. Baraka, "Technology and Ethos," 155–57.

17. Veal, *Dub*; see also Katz, "Dubbing Is a Must"; and Henriques, *Sonic Bodies.*

18. Gilroy, *Black Atlantic.*

19. Hall, *Familiar Stranger.*

20. Hall, "Encoding and Decoding in the Television Discourse," 3; see also Larkin, *Signal and Noise*; Newton, *Paving the Empire Road.*

21. Hendy, "*Caribbean Voices.*"

22. BBC, *Hello! West Indies.*

23. Brathwaite, *History of the Voice*, 87.

24. Hendy, "Caribbean Voices."

25. Bennett, "Colonization in Reverse."

26. Edwards, *After the American Century.*

27. Johnston and Robertson, *BBC World Service.*

28. *New York Times*, "Balfour Unveils Bush House Statue."

29. The United Nations has a radio broadcast network, established in 1946, which plays a similar function today.

30. Ismailov, Gillespie, and Aslanyan, *Tales from Bush House*, 139.

31. Derisking is the practice of taking steps to avoid further or future financial loss by terminating business dealings with specific clients or in localized markets.

32. See Hilmes, *Network Nations.*

33. According to Bloomberg, "Kato Kagaku Co. Ltd. is a diversified products manufacturer. The Company manufactures and markets corn starch, chemically modified starch, animal feed, fertilizer, corn syrup, high fruit corn syrup, dextrose monohydrate and crystalline fructose. Kato Kagaku also operates hotels, commercial buildings, apartment buildings and warehouses."

34. See Hendy, *BBC*; Briggs, *History of Broadcasting in the United Kingdom*; and Scannell, *Social History of British Broadcasting.*

35. Grant, *Bageye at the Wheel.*

36. Grant, *I'm Black So You Don't Have to Be.*

37. BBC, "World Service Staff Bid Farewell to Iconic Bush House."

References

Baraka, Imamu Amiri [LeRoi Jones]. "Technology and Ethos." In *Raise Rage Rays Raze: Essays since 1965*, 155–58. New York: Vintage, 1972.

Baucom, Ian. "Frantz Fanon's Radio: Solidarity, Diaspora, and the Tactics of Listening." *Contemporary Literature* 42, no. 1 (2001): 15–49.

BBC. *Hello! West Indies / West Indies Calling.* London: Ministry of Information and Paul Rotha Productions, Imperial War Museum, 1943.

BBC. "World Service Staff Bid Farewell to Iconic Bush House." February 29, 2012. https://www.bbc.com/news/av/world-radio-and-tv-17198031.

Bennett, Louise. "Colonization in Reverse." In *Jamaica Labrish*, 182–83. Kingston: Sangsters, 1966.

Bloomberg. "Kato Kagaku Ko Ltd." https://www.bloomberg.com/profile/company /6646476Z:JP.

Brathwaite, Edward Kamau. *History of the Voice: The Development of Nation Language in Anglophone Caribbean Poetry.* London: New Beacon, 1984.

Briggs, Asa. *The History of Broadcasting in the United Kingdom*. Oxford: Oxford University Press, 1995.

Connole, Joseph. "The Comanche Code Talkers of World War II: 'A Nation Whose Language You Will Not Understand.'" *Whispering Wind* 40, no. 5 (2012): 21–26.

Denning, Michael. *Noise Uprising: The Audiopolitics of a World Musical Revolution*. London: Verso, 2015.

Edwards, Brian. *After the American Century: The Ends of U.S. Culture in the Middle East*. New York: Columbia University Press, 2015.

Fanon, Frantz. *A Dying Colonialism*. New York: Grove.

Fanon, Frantz. "This Is the Voice of Algeria." In *A Dying Colonialism*, translated by Haakon Chevalier, 69–98. 1959; repr., New York: Grove Press, 1967.

Gilroy, Paul. *The Black Atlantic: Modernity and Double Consciousness*. Cambridge, MA: Harvard University Press, 1993.

Goffe, Tao Leigh. "Bigger than the Sound: The Jamaican Chinese Infrastructures of Reggae." *small axe* 24, no. 3 (2020): 97–127.

Grant, Colin. *Bageye at the Wheel: A 1970s Childhood in Suburbia*. London: Vintage, 2013.

Grant, Colin. *I'm Black So You Don't Have to Be: A Memoir of in Eight Lives*. London: Vintage 2023.

Hall, Stuart. "Encoding and Decoding in the Television Discourse." In *CCCS Selected Working Papers*, edited by Ann Gray et al., 2:386–98. 1973; repr., Birmingham: University of Birmingham, 2007.

Hall, Stuart. *Familiar Stranger: A Life between Two Islands*. Durham, NC: Duke University Press, 2017.

Hendy, David. *The BBC: A People's History*. London: Profile, 2022.

Hendy, David. "Caribbean Voices." In *History of the BBC*. https://www.bbc.com/historyofthebbc/100-voices/people-nation-empire/caribbean-voices/.

Henriques, Julian. *Sonic Bodies: Reggae Sound Systems, Performance Techniques, and Ways of Knowing*. New York: Continuum, 2011.

Hilmes, Michele. *Network Nations: A Transnational History of British and American Broadcasting*. London: Routledge, 2012.

Ismailov, Hamid, Marie Gillespie, and Anna Aslanyan. *Tales from Bush House*. London: Hertfordshire Press, 2012.

Johnston, Gordon, and Emma Robertson. *BBC World Service: Overseas Broadcasting 1932–2018*. London: Palgrave Macmillan, 2019.

Katz, David. "Dubbing Is a Must: A Beginner's Guide to Jamaica's Most Influential Genre." *Fact*, April 16, 2014. https://www.factmag.com/2014/04/16/dubbing-is-a-must-a-beginners-guide-to-jamaicas-most-influential-genre/.

Larkin, Brian. *Signal and Noise: Media, Infrastructure, and Urban Culture in Nigeria*. Durham, NC: Duke University Press, 2008.

Mignolo, Walter. *The Darker Side of Western Modernity: Global Futures, Decolonial Options*. Durham, NC: Duke University Press, 2011.

Newton, Darrell. *Paving the Empire Road: BBC Television and Black Britons*. Manchester, UK: Manchester University Press, 2011.

New York Times. "Balfour Unveils Bush House Statue; Americans in London and Prominent Britishers Attend Independence Day Ceremony." July 5, 1925. https://www.nytimes.com/1925/07/05/archives/balfour-unveils-bush-house-statue-americans-in-london-and-prominent.html.

Orwell, George. "Poetry and the Microphone." In *My Country Right or Left*, vol. 2

of *The Collected Essays, Journalism, and Letters of George Orwell*, edited by Sonia Orwell and Ian Angus, 1–5–110. 1945; repr., London: Penguin, 1970.

Quijano, Anibal. "Coloniality of Power, Eurocentrism, and Latin America." *Nepantla: Views from South* 1, no. 3 (2000): 533–80.

Satia, Priya. "War, Wireless, and Empire: Marconi and the British Warfare State, 1896–1903." *Technology and Culture* 51, no. 4 (2010): 829–53.

Scannell, Paddy. *A Social History of British Broadcasting*. London: Wiley Blackwell, 1991.

Thomas, Deborah A. "Displacements: The Jamaican 1950s." *small axe*, no. 63: 53–64.

Veal, Michael. *Dub. Soundscapes and Shattered Songs in Jamaican Reggae*. Middletown, CT: Wesleyan University Press, 2007.

Williams, Raymond, and Michael Orrom. *Preface to Film* London: Film Drama, 1954.

"This Is Not the Sound of the Asian Underground"

On the Borders of Britain's New Asian Kool, 1997–2000

Tariq Jazeel

On September 25, 2000, the British musician Farook Shamsher and his band Joi released their second album, *We Are Three*. The penultimate song on this CD album is called "Tacadin." It is a high-energy dance track that builds progressively, layering percussive beats, a heavy bassline, keyboards, vocal synths, and a Bansuri flute riff, as well as a Bengali vocal sample. The track's rhythmic continuity sustains a frenetic energy that, like other British Asian dance musics that had begun to emerge from the mid-1990s, was aimed in no uncertain terms at the dance floor. In the track's opening few seconds a male voice declares, "This is not the sound of the Asian Underground. This is music." At forty-four seconds the refrain is repeated, followed by the assertion, "We are here. We are here to stay. To stay to fight. To fight for a better living. For peace, on this Earth." The voice has no discernible Bengali intonation. Accentually, it is marked simply as British, more specifically as a voice from the Southeast of England, probably London. And although the community it speaks for seems to bypass the nation-state by appealing to "peace, on this *Earth*," the refrain makes reference to what, at the time, was a very British music genre: Asian Underground. But it does so through disidentification, disavowal, negation.

"Tacadin" was never released as a single, and the album enjoyed only moderate success. Nonetheless, the song's refrain speaks to a broader tension that pulled at the seams of British Asian dance music in the late 1990s and early 2000s. Following the commercial, critical, and cultural success of the pioneering British Asian musician Talvin Singh, Asian Underground,

Social Text 162 · Vol. 43 No. 1 · March 2025
DOI 10.1215/01642472-11573367 © 2025 Duke University Press

or "New Asian Kool," had emerged in 1997 as an amorphous yet distinctive musical and subcultural genre that gave both voice and visibility to a form of British Asian sonic modernity. This new and emergent genre comprised music that was being produced by British-born South Asians who had grown up in contexts replete with South Asian cultural and familial influences, mostly Indian and Bengali, while they were immersed in a popular culture that through the 1980s and 1990s still had trouble parsing its avowed Britishness from its pervasive whiteness. Musicians like Singh, Nitin Sawhney, Sam Zaman (aka State of Bengal), Osmani Soundz, the Asian Dub Foundation (ADF), Fun>Da>Mental, and a handful of others in the middle to late 1990s were inspired by influences as diverse as dance music, drum and bass, breakbeat, rock and pop, and hip-hop and rap, but equally by classical South Asian instrumentation and musical aesthetics. Asian Underground music quickly and crudely became known as a musical fusion of east and west, and it rapidly gained notoriety and popularity in the mid-1990s through the 2000s. Unlike the Bhangra culture that preceded and ran alongside it, the Asian Underground scene resonated with both white and Asian clubbers and listeners, as well as mainstream and commercial radio DJs. It offered the possibility of a performance culture able to cross over, to speak between brown and white experiences of Britishness, and in so doing, open what Paul Gilroy refers to as a self-consciously postcolonial space in which the affirmation of difference pointed to a more pluralistic conception of nationality and, at least as Joi would have it, beyond that to the humanistic and utopian promise of the transcendence of nationality.[1]

As I suggest in this article, to try to pin down what musically constituted the Asian Underground as a genre is a futile task, such was the diversity of the music for which the term was mobilized. Instead, what interests me here are the tensions around that elusive musical object-genre, and what identifying or disidentifying with it did, not just for British Asian musical/expressive culture but also for the emergence of alternative public spheres where South Asianness became a visible and constitutive element of contemporary Britishness and British modernity. This is, therefore, an article about the performative and iterative instantiation and disavowal of a music genre, its border crossings, and the politics of these in the recent historical context of British Asian belonging. It is, moreover, an article that explores the work that a particular generic categorization, Asian Underground, and importantly, its negation did for the articulation of Britishness with expressive popular culture in Britain at the turn of the last century. Joi's performative disavowal of Asian Underground music spoke to a broader discomfort with the genre among second-generation British–South Asian musicians in the early 2000s, particularly with respect to the genre's constraining implications for musicians who, at the time, wanted

the right to simply be popular musicians alongside and on an equal footing with their white counterparts. Nonetheless, they and the industry they
navigated were also well aware of the important strategic work that the
Asian Underground did to carve a space from where these disavowals and
expressions could be not just voiced but also heard. It is not insignificant,
in other words, that Joi's *We Are Three* album was described by one US
reviewer as "a powerful message *from the Asian Underground music scene*"
(my emphasis), or that in the same review the band's UK label manager
describe Joi as "seminal players *in the Asian Underground rave/club scene*
over the past 10 to 15 years."[2] If Joi felt constrained enough by the term
to explicitly disidentify with Asian Underground (conceived as a genre)
in their track "Tacadin," as a music act they were at the same time com-
mercially enabled by it. It is this tension that I explore. That is to say, this
article pushes at the tensions between the stultification of cultural reification, or essentialization, via generic classification, on the one hand, and on
the other, the kind of presencing, in both cultural and political economic
terms, that the reification of a genre as mutable and contested as Asian
Underground enabled. As I argue, the spatial politics of the *underground*
as a metaphor is key to unraveling the politics of British Asian dance
music in any discussion of the contested contours of postcolonial and
postmillennial Britishness. Asian Underground, I stress here, comprised
a contested, ideological, and ephemeral space from which British-born
South Asians at the turn of the millennium were able to articulate, however problematically, with the nation-state.

From a Critique of Exotica to Musical Mediation

The broader project from which this article emerges is simply an attempt to
write the recent history of this music scene, critically exploring the role the
Asian Underground played in the emergence of new understandings, experiences, and spatialities of Britishness, antiracism, and multiculturalism.
The project also looks to delineate and underscore the lasting memories,
legacies, and failures of the music's postcolonial politics. However, this
music scene and its precedents have been well covered critically already.
Talvin Singh and Nitin Sawhney, in particular, have been subjects of useful if, as I want to suggest, problematic Marxist critiques from anthro-
pologists and sociologists writing in the early 2000s. My own work here
needs to be positioned in the wake of that critique.

It is important to stress that the critiques to which I refer were con-
temporaneous with the emergence of the Asian Underground music scene,
just as they made a clear politico-intellectual distinction between what
this body of work perceived was two divergent strands within the new
Asian dance music. John Hutnyk, Sanjay Sharma, and others were quick

to celebrate the dissident politics that was clear and explicit in the community and protest music epitomized by two bands in particular.[3] First, Fun>Da>Mental, formed in 1991, was a hip-hop break-beat outfit whose music drew variously on Qawwali sounds, Islamic chants, Hindi film samples, and a range of other Islamic and South Asian influences. The group was led by Aki Nawaz, who cofounded Nation Records in 1988 in response to the difficulty he had securing record deals with both major and independent labels in the late 1980s. Second, the Asian Dub Foundation (ADF) formed in 1993, described by sociologist Sharma as "a collective arising from a community music project."[4] ADF's style combined rap, dub, hardcore, and distorted rock guitars with South Asian instrumentations. What characterized the music emerging from the likes of Fun>Da>Mental and ADF through the 1990s and into the 2000s was its explicit antiracist political intent and consciousness. While Fun>Da>Mental's music and iconography articulated a militant Islamic-influenced, pro-Black antiracist politics, ADF focused on contemporary antiracist campaigns they considered to be "real areas for investigation"—abbreviated in the title of their 1998 album, *R.A.F.I.'s Revenge*, a high-octane musical statement in mid- to late-1990s Britain. The influence of hardcore, jungle, and rap was sonically and energetically evident to all who would listen to this East London quintet of young British Asian musicians, and *R.A.F.I.'s Revenge* was meant to signify the band's desire to address politics through their music. This explicit political intent was evident in tracks like "Naxalite," a homage to the Naxalite movement in India, and "Free Satpal Ram," a protest rap focusing attention on the incarceration of Satpal Ram, a British Bengali man who in 1987 was charged and convicted of killing a man during a fight. Ram and a friend were defending themselves from a racist attack by six men in a restaurant in Birmingham. At the time of the album's release, Ram had served eleven years in prison, and would only be let out on parole four years later, in 2002. If the track's energy, discordant guitar riff, and angry lyric spew the ADF's barely contained rage, the song was also a way of bringing this institutional racism and injustice into representation for a broad audience. For these kinds of reasons, Virinder S. Kalra and John Hutnyk, two of the main protagonists in this early body of Marxist anthropological and sociological work on new Asian dance music, described the ADF as "cultural workers who use music to express the frustrations and experiences of young Asian males."[5]

Against this politically conscious New Asian dance music, these scholars counterposed what they saw as a different strand of music from the likes of Talvin Singh, Nitin Sawhney, Joi, State of Bengal, and others. Their less angry, more melodic, and for the most part more instrumental dance and club music constituted a more "sanitized" form of new Asian dance music that, so the critique goes, lent itself to commodification, appropriation, and neo-Orientalizing forms of liberal multicultur-alism.[6] In a searing critique of Asian Underground, Kaushik Bannerjea wrote how "cross cultural music flows, transnational music cultures do not of themselves hijack the power structures of corporate hegemony; or the mobile terrors of white supremacy. Eulogizing Talvin Singh on a Sun-day afternoon at his club in Brick Lane does little to hide white distaste for the large Asian community which actually lives there."[7] Later on in the same article, Bannerjea warns that "finding uncritical pleasure in the *Soundz of the Asian Underground*... is a surefire tactic for denying the angrier sounds of Asian marginality."[8] Maybe. But while these arguments are important and necessary reminders of the perils of the appropriation of Black expressive culture,[9] they would seem to place a certain burden of political representation on British South Asian musicians: If you do not address racism directly in your music, you buttress white supremacy. This is an analysis of music's politics that is at once too sociological, in its dis-regard for the formal qualities and social effects of the musical text, and too focused on a naïve insistence that much Asian Underground music is politically deficient because those musical texts do not explicitly advocate an antiracist politics. As I elaborate below, what interests me instead is the challenge of bringing into focus the music's political effects *as* a form of expressive culture, one inherently linked to modes of spatial production, in particular, the production of new spatialities correspondent with emer-gent modes of postcolonial South Asian Britishness. Take, for example, the 1997 release by the State of Bengal (aka Sam Zaman) titled "Flight IC408," which became something of an anthem in the Asian Under-ground club scene. This track cannot be so easily dismissed as a distrac-tion to South Asian marginality. It begins with an airport boarding gate announcement for "Flight IC408 to Calcutta" (spoken by what we sup-pose is an Indian man, in heavily accented English) and proceeds through seven fizzing minutes of drum and sitar loop, heavy bass, guitar riffs, and a melody plucked by an Indian sarangi or veena, all punctuated by fur-ther flight announcements and the sporadic sounding of a jet engine. This song makes no claims to represent, or protest, any political or racial injus-tice, yet its politics is embedded in its sounding of travel, routes, depar-tures, and arrivals and the incessant mobility and stretched-out belong-ings that cut across static and absolutist imaginations of nationalism and cultural purity. The track's ability to speak to that common, both-here-and-there spatiality—a second-generation South Asian condition—was why it became such a hit on the dance floor in what emerged as the Asian Underground scene. That was its politics.

In this sense, the broader sweep of my project argues that, despite the perils of commodification and exoticization that have been well docu-mented by Bannerjea, Hutnyk, and others, historically there *has* been a progressive politics in the "uncritical pleasure" derived from the Asian Underground that Bannerjea's article impugns. Beyond, then, a somewhat unhelpful polemic narration of the scene, the more salient question that Bannerjea poses in his article is, "What is the symbolic economy of a cultural signifier like the 'Asian Underground'?"[10] This is essentially the question this article tackles. Taking music's ontology seriously, while making no demands that (British Asian dance) music's (antiracist) politics come prepackaged in the music text's symbolic and representational form, proffers answers to Bannerjea's important question that allow us to see how the genre's tensions with which this article began themselves reveal emergent ruptures in the forms of white supremacy that characterized British popular culture in early and mid-1990s.

By referring to music's ontology, I gesture toward a set of conversations in music studies around the challenge of locating music's meaning. Georgina Born is among those who have written on the concept of musical mediation, in so doing mobilizing a relational methodology for locating music's often elusive meaning.[11] Following Lydia Goehr's seminal book *The Imaginary Museum of Musical Works*, Born dispels any sense that it might be easy for us to tell where the locus of music's meaning "really" lies. For Goehr, "thinking about music in terms of works is not straightforward." In other words, Goehr's work alerts us to the pernicious and misguided tendency to regard music as a discrete and divine object, disconnected from the world in which it is both conceived and performed. As Goehr writes,

> Most of us . . . tend to see works as objectified expressions of composers that prior to compositional activity did not exist. We do not treat works as objects made or just put together, like tables or chairs, but as original, unique products of a special, creative activity. We assume, further, that the tonal, rhythmic, and instrumental properties of works are constitutive of structurally integrated wholes that are symbolically represented by composers in scores. Once created, we treat works as existing after their creators have died, and whether or not they are performed or listened to at any given time. We treat them as artefacts existing in the public realm, accessible in principle to anyone who cares to listen to them.[12]

Against this model of the heroic and divine creation of music, as Born puts it, Goehr recognizes "that there is no single privileged location of musical meaning, but that it may be distributed across and configured by the relations between its several mediations."[13] In developing a methodology for engaging music's relational ontology, then, or more specifically, the dynamic production of its meaning, we must look to music's multiple and ongoing mediations, in both the influences and the contextualities that have borne upon its composition and production, but also, and importantly, in attending to what it does in and through the world *as* it is produced and engineered, performed and danced to, commodified and marketed, listened to and engaged with, reviewed and critiqued, and so on. Music's meaning, in other words, is relationally distributed through its social and spatial assemblages. As such, with the benefit of some historical distance (that the likes of Bannerjea, Hutnyk, Kalra, and colleagues did not have), my own project is one that speculates on the value of thinking about what Asian Underground has done as a contested music genre and scene in and through the world, what its emergence and disavowal can tell us about a recent history of British Asian belonging. For, as Tia DeNora has put it, music's semiotic force does not reside within its forms alone.[14] DeNora alerts us to the importance of attending to the articulations between music and social formations and, as such, its world-making capacities.[15] She reminds us that music is coproductive; it is constitutive of social life like other representational forms. As she puts it, "Of interest then is the reflexive problem of how music and its effects are active in social life, and how music comes to afford a variety of resources for the constitution of human agency, the being, feeling, moving and doing of social life."[16] In this sense, then, even those "non-political" (in Bannerjea, Hutnyk, and colleagues' terms) forms of British Asian dance music, which is to say, music that emerged from British Asian musicians who did not consider their work to be protest music, were praxes that in their distributed and relational ontologies effectively reinvented postcolonial life through the prism of heretofore alienated and racialized realities.[17] Their music was kinship making as it was conceived, performed, aired, listened to, and consumed—in short, as it entered and precipitated new kinds of publics and spatialities. The contested spatial politics of Asian Underground was key to this. The next two sections attend to some of the mid- to late-1990s musical practices that assembled the Asian Underground scene, not just as a constitution of the social, but also as a contested spatial production.

The (Asian) Underground

In music studies, the *underground* has tended to denote a particular place in the broader landscape of music making. For Stephen Graham it is a normative category whose analytical value is in its potential to describe music practices outside or at the fringes of the cultural and social mainstream: "The underground can be understood as a distinct zone of cultural activity existing 'below' or 'between,' but permeated by and permeating, the high and low mainstreams."[18] In this configuration, the underground normatively suggests a space of cultural expression with some kind of independence from, as well as relation to, what is a just as normatively conceived mainstream culture. It can be mapped (graphically in Graham's case).[19] It can be described. And, for Graham's particular elaboration on the Extreme Metal scene, it unapologetically denotes a "named thingness."[20] But as I have already stressed, with respect to Asian Underground, the musical forms and praxes that this category emerged to loosely name were so diverse and eclectic that this understanding of the term becomes woefully inadequate. Instead, Asian Underground can best be described as an ephemeral and chimerical genre that from its inception struggled to contain all that it was dubiously made to name. In this sense, what interests me historically and analytically is what work was done by the genre mobilized by the term *Asian Underground* and, importantly, its almost immediate negation, in terms of postcolonial British Asian belonging and the broader contours of British cultural expression.

Asian Underground's emergence as a category, a sign that signified an emergent music genre and its scene, and as popular discourse, can in fact be traced to a cluster of music events in the middle to late 1990s, many of which are associated with the musician Talvin Singh, a second-generation British Asian of Indian Punjabi parentage. Born in East London in the early 1970s, at the age of fifteen he undertook the first of many extended visits to Punjab to learn the tabla from his *guruji* and teacher, Pandit Lachhman Singh Seen. In the late 1980s and early 1990s Singh became known in the British and then international music scene as a gifted and original studio and session musician, and in the late 1980s he was a sought-after programmer, producer, and tabla artist, collaborating with such artists as Courtney Pine, Sun Ra, Massive Attack, and later Björk. In 1996, together with Sweety Kapoor, he took a weekly Sunday music night that he initially ran at the Rocket on Holloway road, the bar at London Guildhall University where Singh was studying art history, to a regular Monday night slot at Hoxton Square's Blue Note club in the East End of London. At the time, the Blue Note hosted some of London's most avant-garde and trendy club nights, including James Lavelle's Mo Wax nights, Ninja Tune's Stealth, and Goldie's Metalheadz. Kapoor and Singh's night, called "Anokha," slotted in next to these. Hoxton Square, the location of the Blue Note club, is less than a mile from London's Brick Lane, which from the early 1950s has been a destination for Bengali immigrants and from the early 1990s has been ground zero for targeted forms of culture- and ethnicity-led urban and economic regeneration strategies. By the late 1990s, Brick Lane had become affectionately known, and branded, as "Banglatown," famed for its burgeoning number of Bengali restaurants.[21] That the club night Anokha had spilled into the "cooler," yet geographically proximate, parts of the City of London is not an insignificant part of this story.[22]

Anokha is a pivotal moment and space in the history of the Asian Underground. From 1996 through to 1997, the club night grew in reputation to become one of the hippest places to be seen in London. As one journalist recalled some twenty years later, "Lines would stretch across the block, and the likes of Afrika Bambaataa, Aphex Twin, Squarepusher, and Bjork (some say even the late David Bowie was a fan) walked through the doors, sometimes ending up behind the decks, till the end of the night."[23] It was an experimental club night at which, despite the impromptu star guest appearance, a range of British-born DJs and musicians, mostly with South Asian heritage, performed, played, toasted, and orchestrated an always ethnically eclectic dance floor. The music, scene, and vibe were experimental, convivial, and importantly at the time, uncategorizable because it was so fresh. Just as importantly, in the middle to late 1990s it provided a space where young British-born South Asians (myself included) could mix with an ethnically diverse crowd of others in the broader context of a soundtrack and vibe that was familiar from familial backgrounds steeped in various forms of South Asian music, aesthetics, and cultural expression. If Singh and Kapoor were the authors of this space, they represented a nascent, British-born, second-generation South Asian sonic modernity. As DeNora writes, "If music can affect the shape of social agency, then control over music in social settings is a source of social power; it is an opportunity to structure the parameters of action."[24] For British Asian clubbers, then, these were "parameters of action" in which a sense of ownership and agency was shared, in which kinships were formed not just with one another but with a sense of what it was to be modern, avant-garde or, put differently, what it was to be "cool" in unapologetic shades of brown, black, and white. At the same time, at Anokha clubbers were aware that listening, dancing, and being there were ways to build new worlds, worlds that they did not yet know the shape of, and that they necessarily were only partially aware they were building. Nonetheless, there was a latent feeling that clubbers were part of something, what Raymond Williams might refer to as an "emergent structure of feeling."[25] In the face of the critique of exotica that Bannerjea leveled specifically at Singh's club night,[26] it is important to stress the ethical and political potential embedded in Anokha's aesthetic, in its "vibe."[27]

In 1997, together with Sam Zaman (better known as the performer State of Bengal), Singh put together a twelve-track compilation CD comprising music by artists associated with Anokha. This as-yet-unnamed album would play a major role in establishing Asian Underground as a generic category that subsequently became so contested. According to Adwait Patil, Singh offered the record to the label London Records, whose artist and repertoire manager at the time, DJ and tastemaker Pete Tong, passed on the record. Singh's recollection of that rejection is as instructive as it is characteristically generous: "'He said he didn't have a place for the record,' Singh says. 'If there is no place for it, it's the underground. So in a way, Pete Tong kind of blessed the record. Even though he rejected it, he blessed it.'"[28] The album was eventually taken up by Chris Blackwell at Island Records and released later in 1997 by Polygram records with the title *Anokha: Soundz of the Asian Underground*. Marketed by Island's Mango Records, the record's title performatively consolidated the ideological subterranean genre and spatiality it named. If Singh's origin narrative for *Asian Underground* is instructive, then Mango Records' marketing and design for the record mobilized the full cultural potential of a term that was contextually implicated in the late twentieth-century history of British popular culture. The CD's artwork consciously drew from a longer lineage of the underground as a trope umbilically tied to the nation in British pop music history. The CD made a direct visual reference to the Ministry of Defence's red, white, and blue Royal Air Force target symbol, which was co-opted and popularized in the 1970s and 1980s by Britain's Mods (see fig. 1). The back cover of the album is adorned by a picture of a figure we assume to be Singh, his back turned to the camera. He is wearing a white parka coat, the informal uniform of Britain's Mods in the 1980s, emblazoned with a version of the logo whose outer blue circle has been turned to green, thus referencing the Indian national flag as it recalls this symbol's nostalgically British semiotics. The words *Asian Underground* emblazoned just below the symbol do more, therefore, than simply reiterate the album's soon-to-be iconic subtitle. They intertextually gesture to the 1980 anthem "Going Underground" by the poster band of British Mod culture, The Jam, a group with which the RAF symbol had become more or less synonymous from the 1970s and 1980s onward, and a group that Singh often names as one of his early influences growing up in London. The visual and symbolic economy of this album's design, marketing, and narrative was not incidental, then, in the performative presencing of a new generic category.

As Simon Frith reminds us, genre distinctions are central to how record company artist and repertoire departments work: "The marketing and packaging policies . . . that begin the moment an act is signed are themselves determined by genre theories, by accounts of how markets work and what people with tastes for music *like this* want from it."[29] What is both striking and perhaps unsurprising, in this respect, is the symbolic and narrative gesture to a nostalgically British popular cultural discourse that is translated as being at one and the same time Asian, Indian specifically. If Pete Tong at London Records had suggested there was no place for this record, Mango Records' marketing savvy was pulling it into what was in fact a well-established place in recent histories of British popular culture: the underground. In so doing, this was a marketing act that visibly, explicitly, and sonically yoked the prevalent whiteness of British his-

Figure 1. Back cover, *Anokha: Soundz of the Asian Underground* (1997)

tory and symbology at the time together with its residual (and constitutive) South Asian presence.

The evident and productive tension here, of course, is that placing this music scene for which no place could previously be found in the ideological spatiality of the underground was in fact to give it a firmly established place in British music and pop history terms. And by virtue of the fact that the CD, unlike the club night, could now be distributed and consumed nationally, the Asian Underground could theoretically be anywhere: in living rooms, cars, bedrooms, across the nation and beyond. Arguably, then, at the moment of its categorical inception, Asian Underground ceased to in fact be an underground spatiality, if such a thing can normatively exist in popular culture. Anokha, as club night and now in its condensed material instantiation as CD, was beginning to precipitate what David Hesmondhalgh refers to as new "forms of collectivity, not only in co-present situations but across space and time."[30]

In the years immediately following Anokha's success, the Asian Underground gained notoriety and presence within the music and cultural sectors. It did so, however, in ways that more or less immediately pulled at the seams of its semiotically sealed borders as a music genre. Talvin Singh released his first solo album, *OK*, in 1998 to critical acclaim. Though it was firmly tied to, and associated with, the Asian Underground, musically the album was a far more worldly and cosmopolitan composition than any straightforwardly hyphenated *British-Asian* moniker might imply. Singh collaborated with a range of musicians for the album, many Indian, but many European, as well as a group of Okinawan folk singers and the Japanese composer, producer, and flautist Ryuichi Sakamota. The album was shortlisted for, and won, the 1998 Mercury Music prize. Though, at the time of writing, Singh remains the only British Asian musician to have won the Mercury Music prize since its inception in 1992, in a four-year period from 1998 through 2001 the prize's shortlist contained at least one British Asian act each year: ADF and Cornershop in 1998, Talvin Singh and Black Star Liner in 1999, Nitin Sawhney in 2000, and Susheila Raman in 2001. Insofar as the term *Asian Underground* was mobilized to account for the sheer diversity of music produced by acts as different as Cornershop and ADF, this short four-year period was perhaps when the generic category was made to do the most work in mainstream music journalism. This was a historical moment that coincided with the early New Labour years, when Tony Blair's newly elected government was explicit about its attempt to make Britain's ethnic and cultural diversity visible.[31] In this historical context, Britain's field of South Asian creative and cultural production was, at the time, burgeoning.[32] However, this short four-year period was also when Asian Underground began to productively unravel as a meaningful category. As I show, this very unraveling not only has been integral and necessary for the (incomplete) normalization of the British South Asian presence within British popular music but also was arguably written into Asian Underground's semiotic trajectory from the very beginning.

Negating a Scene

This is not the sound of the Asian Underground. This is Music.
—Joi, "Tacadin"

I came up with the name because I was really pissed off. . . . It seemed underground because no one was interested in distributing the record. But a lot of Asian artists now blame me for that label. It became an albatross. I'm going to call the next album Soundz of the Asian Overground.
—Talvin Singh, quoted in Nigel Williamson, "Underground, Overground Asian Sounds," *London Times*, September 3, 1999

Almost as soon as its establishment as a discursive category, Asian Underground was contested by many of those to whom it popularly referred.

In part, and as I have stressed above, this was because the sheer diversity of the music it was being used to corral gave the lie to it being a music "genre" that made any sense beyond its racializing connotations. In other words, beyond skin color there was little, if anything, to connect the music of the ADF and Cornershop. Connected with this, however, among many British Asian artists and producers there was a strong sense not only that the term was symptomatic of the prevailing whiteness of a British music industry that still controlled nonwhite musical production, but also that precisely because of this it was restrictive in musical and artistic terms—the music industry's own glass ceiling, as it were. For example, in an interview in the *Times* in 1999, Black Star Liner frontman Choque Hosein complained that "'part of me groans when people talk about the Asian Underground. . . . At school I was the only Asian guy, and I remember the teacher making me play tambourine and bongos in front of the whole class because I was supposed to have a sense of rhythm. It was such a stereotype. People think all Asians can do is dance and cook.'"[33] Though Hosein pushes against the Asian Underground's racializing containment, he makes his statement in the pages of a national daily broadsheet newspaper in an article that covers in some depth Black Star Liner's recent album launch. Importantly, the article's headline reads, "Nigel Williamson meets Black Star Liner and Joi, the latest bands to emerge from *the British Asian Underground.*"[34] The Asian Underground, as generic category, was both an enabling and a constraining category for British Asian musical modernity, an unwitting strategic essentialism.

Even more forthrightly, in a thoughtful feature in the *Independent* in 1998, writer and art critic Hettie Judah solicited musician Nitin Sawhney's opinion on how the late 1990s fetish for Asian Underground music resonated with a broader commercial appropriation of Asian culture in Britain at the time. Sawhney, one of the United Kingdom's longest-standing and most consistent critics of the tyrannies of musical categorization, was forthright in his opinion: "There is still some kind of colonial arrogance operating in western attitudes towards Asia." In a passage worth quoting in full, he reflects more explicitly on music, making a more direct reference to the implication of Talvin Singh and Asian Underground in these categorical tyrannies:

> "I have this idea for a video," Nitin Sawhney tells me. "This Asian kid walks into an underground club, he's kind of got the spiky hair and the rest of it. There's this Talvin Singh type drum 'n' bass music going on and there's all these underground DJs getting into their stuff.
>
> "The camera goes behind the wall and you see this control room full of these old white people dressed in Victorian gear from the days of the Raj, and they are controlling the turntables and the lighting. You go back to the kid dancing and the queues of people having mehndis painted on their hands

by these old people. They are confused and shocked, but they go out and say to their friend 'check out my mehndi' and stuff. Anyway, at the end, the wall collapses and the white people are staring at the Asian kids and they are all staring at each other, and all the DJs are made of cardboard and collapse. What I'm trying to say is, it's still the same; we are still controlled by the British Empire, no matter what you think you've got. The whole 'Asian underground' thing means nothing."[35]

But these critiques were not just about the postcolonial politics of racialization and white supremacy in the music industry. They were about the struggle to break through and out of the subterranean holding space that the underground implied and, in so doing, to sell records in greater volume than the niche market that Asian Underground prescribed. Sawhney was signed to Outcaste Records, a label established in 1995 whose music manifesto was to increase the visibility of British Asian culture. As the owner and founder of Outcaste Records, Shabs Jobanputra, put it in an interview following Talvin Singh's Mercury Music Prize success in 1999, "Let's face it, . . . we've got to sell records in big numbers. If we live by the tag of Asian underground forever, we will fall down. In 10 years it will be forgotten."[36] The more important challenge was mainstream commercial success, the likes of which were still not routinely available to British Asian music acts in the late 1990s. Sawhney released the album *Beyond Skin* in 1999, and it soon became a critically acclaimed and landmark composition that won a prestigious South Bank Show Award and was shortlisted for the Mercury Music Prize in 2000. He has gone on to enjoy huge commercial and artistic success in a thus-far thirteen-studio-album career that has also seen him collaborate with the BBC Proms, the dancer Akram Khan, and many others in the arts, both nationally and internationally. Sawhney has been one of those British Asian artists who *has* effectively transcended any racialized demarcation, as well as any spatialized subterranean containment. In spatial and metaphorical terms, then, he has successfully navigated the passage from underground to overground. Indeed, Talvin Singh himself was not at all unaware of the broader commercial challenge at stake in the face of the generic reifications that Asian Underground presented. In the album sleeve notes for *Anokha: Soundz of the Asian Underground*, Singh signals the challenge ahead that the genre poses in his description of the track "Heavy Intro," performed by the female vocalist Amar:

> I hear that voice right within. Amar is Asian soul—21st century Asian soul. Ever since tuning in to an Asian radio station & hearing a fully blown voice of a 14 year old singing a R&B tune in Hindi & with the internation [*sic*] & attitude of Indian music, I have been totally inspired. *The voice of the British-Asian Underground. Amar's got the ticket to take it overground.*[37]

A Spatial Politics of the Underground

These tense and at times agonistic debates around Asian Underground as genre were not incidental to the kinds of openings that the likes of Sawhney have enjoyed since the late 1990s. The production of a more capacious British popular music whose Asian influences no longer needed to be marked as Asian emerged from the kinds of contestations highlighted in the previous section. In this respect, to seek to deduce whether Asian Underground was a normatively flawed music genre or a category inadequate to capture what was going on musically at the time would seem to be the wrong type of question to ask. Instead, and building on approaches that regard music's ontology as distributed and inherently relational, I have shown in this article the force of the category itself, what it has helped precipitate, even and especially in its negation. In attending to the discursive life of a category that, by the mid-2000s, was more of a historical legacy than meaningful music genre for music industry and public alike, this article has shown how the genre and its contestations themselves performed social actions.[38] In this sense, we can read Joi's musical proclamation, "This is not the sound of the Asian Underground," for its perlocutionary effect. It matters not, in other words, how we *should* musically categorize "Tacadin"; what matters instead is the track's intentional utterance that helps produce an effect. As Mary Jo Reiff and Anis Bawarshi put it, it is within the interplay and trans-actions between genres or, in this case, at a genre's own blurred borders that the performances and productions of publics take place. Reiff and Bawarshi write in the context of radical genre studies scholarship that has usefully shown not only how public genres are always constituted through dynamic discursive performances but also that that publics speak and inscribe themselves into existence through the very performance and discursive contestation around the meaning, capacity, and limits of a genre.[39] Indeed, genre scholar Charles Bazerman goes further by suggesting that understanding genre as emergent and dynamic performance also helps us understand how individuals write themselves into citizens.[40]

It is therefore in the context of Asian Underground, a genre whose destiny was its categorical demise, that we can read the ADF's claim in 1999 that "We're more British than Oasis."[41] Likewise, in reviewing the ADF's Kentish Forum gig in the *New Musical Express* in 1999, music journalist Steven Wells wrote that the "Asian Dub Foundation are the best British band since The Clash."[42] It is also in this context that we can read a 1999 performance by Nitin Sawhney at none other than London's Victoria and Albert Museum, where one reviewer describes his music as nothing short of "a template for Britain's culturally diverse future."[43] In 1999, these assertions of nonwhite—and specifically Brown—Britishness were bold in a context where there was a considerable and angry public backlash against the Runnymede Trust's Parekh report, which called for a reevaluation of Britain's national narrative and emphasized that British national identity cannot, and should not, be preserved in stasis like some antique piece of furniture. Whether they knew it or not, what Singh, Kapoor, and Island's Mango Records had done with the categorical delineation of Asian Underground was to provide something of a discursive and commercial opening, a tear in the fabric of a British popular music industry that, prior to its existence, could find no place for British Asian music. That the term *Asian Underground* had such a short historical life, with the demise of the genre, should be taken not as its failure but instead as an indication of the considerable work the genre and its contestation have done to prize open the seams of Britishness and what it can mean.

In closing, it is prudent to sound a warning about any uncritical celebration of that brief moment from around 1997 to the turn of the millennium when Asian Underground emerged, only to be almost immediately contested. If, in this article, I have set out the genre's trajectory in that short period in ways that point to the openings in, and expansions of, expressive postcolonial culture and the intuitions of Britishness that it precipitated, then Bannerjea's concerns about the "uncritical pleasure[s]" taken in the Asian Underground scene—pleasures we might also frame, more than two decades on, as somewhat nostalgic—are still worth taking seriously.[44] The perils of exotica, the lure of superficial flirtations with racialized otherness, and the fractal half-life of hyphenated belongings were all part of this historical conjuncture as well. And racism remains. In addition, we should note that in 2006 the BBC's Asian radio network was relaunched nationally, and from that moment onward much emerging British Asian musical talent was siphoned into this segregated airspace. What Asian Underground and its generic contestation achieved, however, was undoubtedly a kind of reinvention of Brown British life. Even though Singh, Sawhney, Joi, and others would never claim to be "cultural workers who use music to express the frustrations and experiences of young Asian males [*sic*],"[45] I have argued here that there was undoubtedly a politics to their music. As I have shown, however, this was a politics made visible only when their music's ontology is properly seen to be distributed across, and positioned in relation to, the ways it was consumed, listened to, grooved to, reviewed, categorized, critiqued, and so on. It was a politics attached to Asian Underground's configuration as not just a music but as a contested scene. In this respect, tracking the trajectory of the Asian Underground as a genre that was doomed from the moment of its inception has aimed at showing how the marginalities of South Asian expressive culture within a late twentieth-century British polity became relatively mainstream, accepted, cool even, and part of what it means to be British today, however problematic that may still be.

Tariq Jazeel teaches geography and race, ethnicity, and postcolonial studies at University College London. His research is positioned at the intersection of postcolonial theory, critical geography, and South Asian studies. He is professor of human geography and former codirector of University College London's Sarah Parker Remond Centre for the Study of Racism and Racialisation. He is author of *Postcolonialism* (2019) and *Sacred Modernity* (2013), coeditor of *Subaltern Geographies* (2019) and *Spatializing Politics* (2009), and is currently writing a book on the history of the Asian Underground scene.

Notes

My thanks to the participants of the Sound Carries workshop at University College London's Institute of Advanced Studies in June 2022 for their comments, and to audiences in the Department of Music at Kings College London and the Department of Geography at Aberystwyth University. Thanks also to Clare Finburgh Delijani, Tom Western, Georgina Born, Emilia Weber, and the two anonymous referees for their comments, and to Marie Buck at *Social Text*.

1. Gilroy, *Small Acts*, 62.
2. Van Vleck, "Real World Releasing Joi in January."
3. See, e.g., Hutnyk, *Critique of Exotica*; and Sharma, Hutnyk, and Sharma, *Dis-orienting Rhythms*.
4. Sharma, "Noisy Asians or 'Asian' Noise?," 47.
5. Kalra and Hutnyk, "Brimful of Agitation, Authenticity, and Appropriation," 351. Though the main musicians on the "scene" were principally male, women were visibly present as clubbers, listeners, and consumers of the music. The late twentieth-century Asian Underground scene would also prove a catalyst for a next generation of musicians and DJs somewhat more evenly mixed in terms of gender. For more on gender and Asian Underground, see Bakrani, *Bhangra and the Asian Underground*.
6. See Hutnyk, *Critique of Exotica*, 3–18; and Kalra and Hutnyk, "Brimful of Agitation, Authenticity, and Appropriation."
7. Bannerjea, "Sounds of Whose Underground?," 65.
8. Bannerjea, "Sounds of Whose Underground?," 76.
9. Gilroy, *Black Atlantic*, 3–4.
10. Bannerjea, "Sounds of Whose Underground?," 71.
11. Born, "On Musical Mediation."
12. Goehr, *Imaginary Museum of Musical Works*, 2.
13. Born, "On Musical Mediation," 9.
14. DeNora, *Music in Everyday Life*, 41.
15. DeNora, *Music in Everyday Life*, 6; also see Born, "On Musical Mediation," 13.
16. DeNora, *Music in Everyday Life*, 45.
17. McKittrick, *Dear Science*, 161.
18. Graham, *Sounds of the Underground*, 8.
19. Graham, *Sounds of the Underground*, 9.
20. Graham, *Sounds of the Underground*, 6.
21. For more on the history of culture-led regeneration at Brick Lane, see Shaw, "Marketing Ethnoscapes as Spaces of Consumption"; and Keith, *After the Cosmopolitan?*
22. On art, regeneration, and gentrification in Hoxton, see Harris, "Art and Gentrification."
23. Patil, "Look Back."

24. DeNora, *Music in Everyday Life*, 20.

25. Williams, *Marxism and Literature*, 128.

26. Bannerjea, "Sounds of Whose Underground?"

27. McKittrick, *Dear Science*, 167.

28. Singh, quoted in Patil, "Look Back."

29. Frith, *Performing Rites*, 76.

30. Hesmondhalgh, *Why Music Matters*, 85.

31. See Power Sayeed, *1997*; and Arday, *Cool Britannia and Multi-ethnic Britain*.

32. For example, Monica Ali's novel *Brick Lane* was published in 2003; the BBC radio and then TV comedy *Goodness Gracious Me* ran from 1996 to 2001, Ayub Khan-Din's movie *East Is East* was released in 1999, and Gurinder Chadha's *Bend It like Beckham* in 2002.

33. Hosein, quoted in Williamson, "Brimful of Asia on the 45," 47.

34. Williamson, "Brimful of Asia on the 45," 47; emphasis added.

35. Sawhney, quoted in Judah, "Real People," 2.

36. Jobanputra, quoted in in Rajan, "Music," 14.

37. Singh, sleeve notes, *Anokha: Soundz of the Asian Underground*; emphasis added.

38. Reiff and Bawarshi, "From Genre Turn to Public Turn," 4.

39. Reiff and Bawarshi, "From Genre Turn to Public Turn," 5–10.

40. Bazerman, "Genre and Identity," 34.

41. ADF, quoted in Williamson, "British Folk Comes of Asia," 40.

42. Wells, "Asian Dub Foundation."

43. Remi, "Pop Review."

44. Bannerjea, "Sounds of Whose Underground?," 76.

45. Kalra and Hutnyk, "Brimful of Agitation, Authenticity, and Appropriation," 351.

References

Arday, Jason. *Cool Britannia and Multi-ethnic Britain: Uncorking the Champagne Supernova*. London: Routledge, 2020.

Bakrani, Falu. *Bhangra and the Asian Underground: South Asian Music and the Politics of Belonging in Britain*. Durham, NC: Duke University Press, 2013.

Bannerjea, Kaushik. "Sounds of Whose Underground? The Fine Tuning of Diaspora in an Age of Mechanical Reproduction." *Theory, Culture, and Society* 17, no. 3 (2000): 64–79.

Bazerman, Charles. "Genre and Identity: Citizenship in the Age of the Internet and the Age of Global Capitalism." In *The Rhetoric and Ideology of Genre: Strategies for Stability and Change*, edited by Richard Coe, Lorelei Lingard, and Tatiana Teslenko, 13–37. Cresskill, NJ: Hampton, 2002.

Born, Georgina. "On Musical Mediation: Ontology, Technology, and Creativity." *Twentieth Century Music* 2, no. 1 (2005): 7–36.

DeNora, Tia. *Music in Everyday Life*. Cambridge: Cambridge University Press, 2000.

Frith, Simon. *Performing Rites: Evaluating Popular Music*. Oxford: Oxford University Press, 1996.

Graham, Stephen. *Sounds of the Underground: A Cultural, Political, and Aesthetic Mapping of Underground and Fringe Music*. Ann Arbor: University of Michigan Press, 2016.

Gilroy, Paul. *The Black Atlantic: Modernity and Double Consciousness*. London: Verso, 1993.

Gilroy, Paul. *Small Acts: Thought on the Politics of Black Cultures*. London: Serpent's Tail, 1993.

Goehr, Lydia. *The Imaginary Museum of Musical Works: An Essay in the Philosophy of Music*. 1992; repr., Oxford: Oxford University Press, 2007.

Harris, Andrew. "Art and Gentrification: Pursuing the Urban Pastoral in Hoxton, London." *Transactions of the Institute of British Geographers* 37, no. 2 (2012): 226–41.

Hesmondhalgh, David. *Why Music Matters*. Oxford: Wiley Blackwell, 2013.

Hutnyk, John. *Critique of Exotica: Music, Politics, and the Culture Industry*. London: Pluto Press, 2000.

Judah, Hettie. "Hands Off Our Culture." *Independent*, December 6, 1998.

Kalra, Virinder S., and John Hutnyk. "Brimful of Agitation, Authenticity, and Appropriation: Madonna's 'Asian Kool.'" *Postcolonial Studies: Culture, Politics, Economy* 1, no. 3 (1998): 339–55.

Keith, Michael. *After the Cosmopolitan? Multicultural Cities and the Future of Racism*. London: Routledge, 2005.

McKittrick, Katherine. *Dear Science, and Other Stories*. Durham, NC: Duke University Press, 2021.

Patil, Adwait. "A Look Back as the Asian Underground Turns Twenty." *NPR: Code Switch*, April 6, 2017. https://www.npr.org/sections/codeswitch/2017/04/06/522 732985/a-look-back-as-the-asian-underground-turns-20.

Power Sayeed, Richard. *1997: The Future That Never Happened*. London: Zed, 2017.

Rajan, Datar. "Mercury Music Prize: Turning the Tablas." *Guardian*, September 10, 1999.

Reiff, Mary Jo, and Anis Bawarshi. "From Genre Turn to Public Turn: Navigating the Intersections of Public Sphere Theory, Genre Theory, and the Performance of Publics." In *Genre and the Performance of Publics*, edited by Mary Jo Reiff and Anis Bawarshi, 3–24. Denver: University Press of Colorado; Salt Lake City: Utah State University Press, 2016.

Remi, Abbas. "Pop Review: Nitin Sawhney at the V&A." *Guardian*, July 12, 1999.

Sharma, Sanjay. "Noisy Asians or 'Asian' Noise?" In Sharma, Hutnyk, and Sharma, *Dis-orienting Rhythms*, 32–60.

Sharma, Sanjay, John Hutnyk, and Ashwani Sharma, eds. *Dis-orienting Rhythms: The Politics of the New Asian Dance Music*. London: Zed, 1996.

Shaw, Stephen J. "Marketing Ethnoscapes as Spaces of Consumption: Banglatown—London's Curry Capital." *Journal of Town and City Management* 1, no. 4 (2011): 381–95.

Singh, Talvin. Sleeve notes. In *Anokha: Soundz of the Asian Underground*. Talvin Singh Presents, CIDM 1120/524 341-2, Omni Records.

Van Vleck, Phillip. "Real World Releasing Joi in January." *Billboard*, December 2000.

Wells, Steven. "Asian Dub Foundation." *New Musical Express*, October 30, 1999.

Williams, Raymond. *Marxism and Literature*. Oxford: Oxford University Press, 1977.

Williamson, Nigel. "Brimful of Asia on the 45 (and LP)." *London Times*, January 29, 1999.

Williamson, Nigel. "British Folk Comes of Asia." *London Times*, July 2, 1999.

Black Folk in English Folk

Les Back and Stevie Back

"Some folksongs are good and some are bad," wrote A. L. Lloyd in his pathbreaking study *The Singing Englishman*, published in 1944. "The best of them are probably the highest points reached by the imagination of the ordinary Englishman . . . as genuine poetry and music produced by East Anglian shepherds and Wiltshire ploughboys and cattle thieves on the Border."[1] "Bert" Lloyd, as he was known, was a gentle, attentive maverick and passionate life-loving Marxist. Born in London in 1908, he was orphaned at fifteen, and his family took the drastic step to send him to Australia as an "assisted migrant," his passage paid by the British Legion. Bert was a laborer and spent his youth working on sheep stations in Australia and subsequently on Antarctic whaling ships.[2] Also a gifted singer, Bert was not your typical midcentury commentator on the folk tradition like the middle-class guardians of the English Folk Dance and Song Society, whose founding fathers were folklorist Cecil Sharp and British composer Ralph Vaughan Williams.[3]

Bert, like many, considered folk music to be a living art form created in communities for communities. He believed that "folk songs belong to the unlettered"[4] and sought to return folk music to the working people who had created it and to the worlds from which these songs emerged, from the fields and the seas to the industrial cities and public houses.[5] We start this article with Bert Lloyd to open a discussion about how to understand English folk music and the way it has been narrowly defined traditionally. Through exploring contemporary folk artists of color, we want to investigate how the contours of the folk tradition are being expanded. We argue that thinking critically about folk music provides useful insights into how music as a living culture and unfolding tradition is understood. The history of English folk music provides a useful example to see how

Social Text 162 · Vol. 43 No. 1 · March 2025
DOI 10.1215/01642472-11573380 © 2025 Duke University Press

processes of exclusion operate in the way folk is defined, who it refers to, and to whom it belongs.

Despite the nation's history of multiculture and empire, until recently Black music in England has rarely been regarded as folk music. There are, however, lots of contemporary artists of color who are reclaiming the English folk tradition, challenging it, modernizing it, and making it their own. Artists like Angeline Morrison, who released an album in 2022 called *Sorrow Songs*, are using the form of English folk song to tell the stories of Black British experience.[6] Others, such as British Caribbean DJ, producer, and writer Zakia Sewell, who became entranced by English folk music as a teenager through a love of Pentangle, are exploring the degree to which the ancient, mythical land evoked in so many folk songs, symbols, and stories can truly admit Black people. We focus our argument on a series of conversations with Angeline Morrison and musician Hakeem ("Hak") Baker, who uses the term *folk* to define his own genre of music: "guv'nor folk," or G-folk.

Before exploring these contemporary voices in more detail, we want to confront the issue of why this matters within the contemporary debates about nation, race, and belonging in the United Kingdom. "Cultural diversity" in the United Kingdom is largely the product of mid-twentieth-century immigration of former colonial citizens to the "Motherland." English racism in this context has often focused on defense of a "way of life" that is threatened by the import of difference from outside. At the same time, as Paul Gilroy has pointed out, "the relationship of England to itself, with its unique cultural traditions and the habits of common life have been so mystified and distorted by the mentalities of empire, militarism and postcolonial nationalism that the country's 'ecology of belonging' can be distinguished by a profound sense of cultural deficit."[7] Gilroy points out that myths of English white supremacy and nationalism, whether in the political mainstream or on the extreme margins, are defined by bellicosity and wartime spirit rather than an attention to folk culture or music. The predominant key of racial nationalism is the melancholic tone of a loss of cultural homogeneity, or of being "ethnically cleansed," and being made uncomfortable by unwanted difference. In this context, Gilroy concludes, "today there is something deeply transgressive about accepting the 'alien' presence of these 'postcolonial' figures in England's deepest folk-cultural recesses, the repository of its ancient lore, its vernacular historical commentaries and its traumatized, reflexive class consciousness, urban and rural."[8] Perhaps it is surprising that the controversies around English folk music documented in what follows provide us with a clue concerning the changing politics of race, nation, and belonging in England. This is what we feel is at stake in the music: the fate of England's past and future.

First we look at how English folk music has been understood and specifically how Black folk have been positioned in relation to English folk. To do this we need to return to the 1930s and the period just prior to the folk revival of the 1960s.

White Folk and Black Albion

In 1967 Bert Lloyd published *Folk Song in England*, which has become a classic history of the folk tradition in England, the culmination of twenty-five years of collecting and collating songs. During the late thirties and forties he was a journalist with an ear for ordinary life and a poetic turn of phrase. He wrote for the *Left Review* and for the *Daily Worker*'s British edition, the *Listener* and collaborated with photographer Bert Harvey for the *Picture Post*. Here is an example from a *Picture Post* article on working-class community life in the Elephant and Castle district of South London: "Its voice has the rasp of trams, trains, trucks. Its eyes have the blaze of street-stalls, eel-stands, pin-table arcades and chestnut cans. Its anatomy is decked with sooty bricks, cast iron spikes and the marble pillars of pubs. Its heart is that of its people—kind as a housewife, rough as a worker, busy as a tradesman, wide as a wide-boy."[9] The vividness of this description reads like a folk song: it captures the unfolding of community life. His journalism, which might be read as a proto-ethnographic social documentary, shaped his understanding of the contexts in which musical traditions are best situated. Lloyd was constantly suspicious of fossilizing culture or conferring folk music to the past, or what Raymond Williams referred to as "fixed forms."[10] In 1939 Marxist historian Leslie Morton took Lloyd to a song school at Eel's Foot Pub in Eastbridge in Suffolk. Morton was another independent scholar who wrote *A People's History of England*, which was a strong influence on Lloyd.[11] Out of that visit, before the outbreak of World War II, came a historic BBC broadcast. By this time Lloyd was producing programs, and he returned with recording equipment to record "The Foggy Dew," "The Blackbird," "Indian Lass," "Poor Man's Heaven," "Little Pigs," "There Was a Farmer in Cheshire," and "Pleasant and Delightful," plus a concertina solo called "Jack's the Boy." The resulting program was broadcast on July 21, 1939.[12]

Morton and Lloyd were intellectuals without a university and deeply suspicious of the trappings of academia and scholarly poses. We return later to why their ideas still matter, but as a starting point, these writers demonstrate both a radical opportunity for the cultural analysis of folk music and a deep limitation concerning their range and scope. Lloyd writes in *Folk Song in England*, "Only a moribund tradition is *dominated* by the past; a living tradition is constantly sprouting new leaves on old wood and sometimes quite suddenly the bush is ablaze with blossom of a novel shade."[13] However, those novel shades were not extended to the emerging music forms that were being made within the Black communities that had been established literally on his doorstep—they were simply not admissible as forms of English folk music.

Lloyd lived in Greenwich at 16 Crooms Hill for more than thirty years, until his death in September 1982. His home was an archive and a meeting place. Many trips were planned here to document folk music in Hungary, Albania, or Transylvania. The journey from London Bridge to Greenwich is portrayed poignantly in Barry Gavin's posthumous 1983 film *The Singing Englishman: A Portrait of A. L. Lloyd*. It was a journey that Lloyd would have made on countless occasions on his research trips. It is a telling clue that he would have literally passed over the railway arches where countless reggae sound systems had their "lock ups" underneath the train tracks in Deptford and New Cross, which went unnoticed by the great folklorist. Through the vehicle of the reggae sound system, collectives of Black Londoner DJs and MCs were making their own folk culture, developing their own unique voice, and documenting their experience in England. This world is portrayed vividly in Franco Rosso's 1981 film *Babylon*. What we want to stress is the gap symbolized in this fable: the gap between an English folk tradition that is defined by whiteness literally passing overhead, and a Black tradition of music making and alternative assembly unfolding underground.[14]

This is not simply the story of Black and white music tracks that did not cross. Rather, it is a matter of how these histories are written and the attentiveness of the folklorists. These untenured historians colluded, perhaps unwittingly, with a definition of English folk music that presumed—explicitly or tacitly—that it belonged to white people. The admissions of open racism in Cecil Sharp's diaries that record his experience of traveling in the United States between 1916 and 1918 shows a clear alignment of race, culture, and music. After a visit to a friend in New York, on Sunday, December 8, 1918, he wrote, "Had tea and afterwards went round to Glenn to say good bye. He resented my dubbing the negroes as of a lower race & maintained it was a mere lack of education etc! Walked round to the Gilmans dined there and alas! said farewell with as little ceremony as possible, but I felt very sad walking home."[15]

Sharp made song-collecting expeditions through the Appalachian mountains in Virginia, North Carolina, Kentucky, and Tennessee with assistant Maud Karpeles, often on foot through rough terrain in search of music of English and Scottish heritage. Sharp willfully ignored the Black population of Appalachians, sometimes turning back disappointedly when he encountered them. Sharp recorded only two songs of English heritage from Black singers, including the popular ballad "Barbara Allen." It seems clear, though, that in large part Sharp felt that English song did not, or could not, belong to Black Appalachians in so-called negro settlements, despite the fact that these songs and tunes were played and sung in Black communities and there was two-way traffic of musical influences across the color line.

Gilroy has commented that there are many examples of Black performers in English folk music, or what he has referred to as a "shadow history of English folk."[16] For Gilroy this includes singer and autoharpist Dorris Henderson, who was born in Florida, the daughter of an African American clergyman. She performed and recorded with English folk guitarist John Renbourn, producing two albums: *There You Go* (1965) and *Watch the Stars* (1967). Another key figure in Gilroy's shadow history is Nadia Evadne Cattouse, who was born in British Honduras and who came to Scotland in 1943 during World War II as a volunteer and was trained in Edinburgh as a signal operator. She subsequently attended teacher training college in Glasgow before returning to British Honduras. She returned to the United Kingdom in 1951 to study social sciences at the London School of Economics and later became an actor (often appearing with Guyanese actor Cy Grant) and a singer. Cattouse was associated with Ewan MacColl, who was at the heart of the sixties folk revival. She made two folk albums: the eponymous *Nadia Cattouse* (1966) and *Earth Mother* (1969). The second of these features a tune called "Bermondsey" that neatly connects with a thread in our argument. The song might have been a musical accompaniment to Bert Lloyd's *Picture Post* community portrait of neighboring Elephant and Castle mentioned earlier. Recording live at the Edinburgh Festival in 1969, she sings in her Honduran accent of London life: "It's nighttime in Bermondsey. The tide is turning now on barges in Bermondsey. The waters laps their bows. And on London Bridge young lovers shiver and gaze at the lamplight in the river."[17] The song not only paints vivid portraits of housewives and pensioners, barrow boys and dockers, doctors and shivering young lovers but also names the iconic features of the local landscape, such as St. Saviour's Church, Guy's Hospital, and Southwark Cathedral. We do not know how she came to learn it, but the song—actually titled "Nighttime in Bermondsey"—was written by Roy Meadow, who was a first-year house officer at Guy's Hospital, and it was part of the finale for the 1961 Guy's Hospital residents' play *Drug Fever*. The lyrics of the song reference "Caleb" and "Diplock," which were two children's wards in Hunts House in Guy's Hospital in the 1950s, when parents were not allowed to visit children at night, which adds to the haunting loneliness of this London ballad.

Black singers like Nadia Cattouse and Dorris Henderson were giving voice to English folk songs from within the culture, thus challenging any racially coded understanding of these musical traditions. This was also true of American Gospel singer Paul Robeson's contact and connection with working-class communities in Britain during the 1930s. Jeff Sparrow in his biography of Robeson described a letter he received from a cotton spinner during one of his tours: "This man said he understood my singing, for while my father was working as a slave, his own father was working as a wage slave in the mills of Manchester."[18] In Sparrow's account, it was in Wales where Robeson developed a sense of shared experience with the British working-class movement and culture. He developed a relationship with Welsh miners and, in 1934, upon hearing the news of the Gresford colliery disaster, gave a concert in Wrexham and donated to the families of the men who lost their lives. In an interview with a local newspaper, Robeson told the reporter that he had come to see himself as a folk singer rather than a classical one and that he was now devoted to what he called "the world body of folk music."[19]

There have been other, more recent attempts to make the idea of folk music in the United Kingdom more inclusive. In 1999 journalist Nigel Williamson wrote a *London Times* review of the London-based Arts Worldwide Bangladesh Festival asking whether headliners Asian Dub Foundation (ADF) might be the "new face of British Folk." ADF, formed in Tower Hamlets in 1993, is perhaps the best early example of the bridging of diasporic South Asian and African Caribbean musical traditions with sound system culture at its center. ADF emerged from the London-based organization Community Music in Farringdon and the Community Music House, an organization/music laboratory for established and aspiring musicians where bassist Aniruddha Das—"Dr Das"—taught music technology. Das, who describes himself as a "Hindi, British, Asian, English, Bengali European," teamed up with one of his students, rapper Deeder Zaman, and civil rights worker John Pandit—"DJ Pandit G"—who describes himself as a "half Irish Asian Scot," to form a sound system to play at antiracist gigs. Responding to Williamson's rhetorical questions, Das replied, "That's spot-on, folk music is about a sense of community. It binds people. And it was always a vehicle for social commentary. That's what we do. And our music draws from everything there is to be heard in modern Britain. It doesn't recognise genres. In 50 years people will look back on our samples and think 'how quaintly old-fashioned.' Of course it's folk music."[20]

However, in the early part of the twenty-first century, folk music was opportunistically appropriated by the British National Party (BNP) as a pinnacle of white nativism. Several prominent folk musicians found that their music was being sold via the party's online shop as a recruitment medium and fundraiser.[21] Then-leader of the BNP Nick Griffin was a self-professed fan of singing English folk music in pubs, with a penchant for fiddle tunes. He himself penned lyrics for an album of "patriotic" songs titled *West Wind*, released in 2007, and is a self-proclaimed fan of the contemporary folk artists Eliza Carthy and Kate Rusby, much to their discomfort.[22] The 2009 edition of the BNP's *Activists and Organ-*

iser's Handbook sets out a strategy to target folk music events like the Padstow Obby Oss and the Whittlesey Straw Bear Festival for recruitment drives. Caroline Lucas, in her analysis of this political moment, writes, "Folk music's place within the discourse of the BNP brings together two of their key political narratives. Firstly, the imagining of the English as an 'indigenous' community, marked as explicitly white, and secondly, the notion that this identity is either lost or threatened."[23]

A collective of some of the most established UK folk artists rallied under the banner of "Folk Against Fascism" to resist this white nationalist attempt to insinuate itself into the folk community and lay claim to the musical tradition. A double album released in 2010 featured Chris Woods, Jon Boden, The Unthanks, Billy Bragg, Damien Dempsey, Andy Cutting, Kris Drever, June Tabor, and Belshazzar's Feast, its cover placing the *Folk Against Fascism* slogan echoing the famous image of American folk singer Woody Guthrie's guitar that had "This Machine Kills Fascists" written on it. In the album's liner notes, Tim Chipping wrote with humorous defiance, "We will not take this sitting down with a finger in our ear. Folk songs are paradoxically rooted and rootless. . . . The British folk tradition is no more a 'pure' representation of Britain's past than the British people can claim to be a 'pure' race. We are a mongrel nation and proud of it."[24] We return to the association of folk with a mongrel national tradition later in this discussion, but here it is important to note that it is being deployed as a mechanism to admit difference to challenge racism.

Some of the songs on the album directly address the politics of race and nation, such as Woods's "Spitfires," which was written as a direct response to a BNP leaflet bearing the image of a Spitfire that came through his letterbox. Bragg's tune with Imagined Village, "English, Half English Meets John Barleycorn," is a celebration of the malleability of culture, as is the way in most people's lives, with images of "veggie curry" being turned into "bubble and squeak." The song offers playful reminders that the icons of English culture have their origins oversees, such as the three lions on the England football shirt and the fact that Saint George was born in Lebanon. The Unthanks's "Nobody Knew She Was There" deals with the loss of an individual sense of self, rather than a collective national identity. Dempsey's song "Colony" addresses the meaning of imperial greed, as seen from the vantage point of the colonized. For most of the artists on this double CD (with thirty-one tracks), it seems the act of lending their music to the project is a statement of solidarity and alignment with the Folk Against Fascism position in the folk tradition.

Perhaps strangely, English folk music, which is often derided or ignored in popular media, has become a particularly interesting cultural and creative space where the terms and meaning of Englishness and its racial coding are being openly contested and struggled with.

The cultural tactics used by antiracist folk artists stress the collaboration and combination of music traditions. Caroline Lucas points out that a good example of this is Imagined Village, whose first album, *Imagined Village Project* (2007), presents "an ambitious reinvention of the English folk tradition." This release includes dub poet Benjamin Zephaniah's "Tam Lyn (Retold)," which also features a vocal from Eliza Carthy. Based loosely on "Tam Lin," a ballad originating from the Scottish borders about a man's rescue from the Queen of the Fairies by his true love. In Zephaniah's retelling, "Tam Lyn" is an "illegal alien" and refugee from war who is saved from the immigration police. In this retelling, Tam's redemption is in multicultural love, which gifts him the "respectable" status of legal citizenship and a child that "grows up to be a club DJ." While not to deny the playful joy in the music, the question remains, What kinds of cultural politics result from these attempts to redefine the folk tradition? Lucas concludes that, while these responses within the folk movement "draw attention to the fluid plurality of culture and the creative potential of musical cooperation, it still manages to (re)produce the whiteness of folk (and hence Englishness) reifying its position in relation to the otherness of the non-white/folk elements."[25]

We argue that artists of color operating inside the English folk tradition are transforming and challenging the whiteness of English folk referred to above. The release in 2022 of Angeline Morrison's collection *The Sorrow Songs: Folk Songs of Black British Experience*, produced by Eliza Carthy, signaled an important shift, as this collection of songs, performed in a traditional English folk style, addresses the legacy of the historic presence of the African diaspora at the center of the folk tradition. The album's title is a reference to the final chapter of W. E. B. Du Bois's 1903 classic of African American literature *The Souls of Black Folk*, titled "Of Sorrow Songs,"[26] which Morrison reread after the killing of George Floyd in 2020.

Morrison (see fig. 1) explained the origin of the idea for the album in an interview:

> In 2020, in the wake of the horrific murder of George Floyd, I reread *The Souls of Black Folk*. And I was really struck. I was really, really struck by the musicality of the text, first of all, and second of all, . . . by what Du Bois had to say about the body of folk songs that African American people have, that the enslaved Africans in America and their descendants have, they've got this body of folk songs. . . . They contain coded messages and . . . stories, and they're just . . . such a hugely important cultural storehouse. . . . And I thought to myself, we now know, without doubt, that we have a historic black presence in Britain. . . . Where is the equivalent body of folk songs by the black people in Britain? I like to use expression Black Albion. . . . Where are the songs of old Black Albion?[27]

Figure 1. Angeline Morrison, 2023. Photograph by Nick Duffy.

The songs on the album are an attempt to give the ghosts of Black Albion a cultural space for a hospitable memory.

The album, with "Unknown African Boy (Died 1830)" as the opening ballad, takes inspiration from a nineteenth-century newspaper story listing flotsam of a shipwreck that washed up on a beach on the Isles of Scilly. The article lists the washed-up cargo of imperial plunder: palm oil, several hundred elephant tusks, a box of silver dollars, two boxes of gold dust, and an unknown "West African boy," estimated around eight years of age. The song is delivered in the voice of the boy's mother, who sings, "O my brown arms, they are sad and empty, O where, o where is my little son? He's stolen away by English slavers, With a cudgel blow, and a pointed gun." What the collection does so powerfully is demonstrate the centrality of the transatlantic trade to the British social formation and Englishness specifically. The final song on the album, "Slave No More," based on the inscription on a grave where master and slave are buried side by side, presses this home. Martin Carthy, an icon of the English folk movement and father of Eliza Carthy, reads the gravestone inscription as part of this song adding an eerie coda to the album.

In an interview Morrison explained her intention:

> I wanted to find stories of...real Black people, real people of color in Britain...present in the land. And then...compose songs about them, which would be sonically similar to the traditional or folk or ancient music of these islands. And which...would not feel out of place in folk clubs or in any other places where British folk songs are sung. I just wanted to give a gift to these ancestors, and also to give a gift to the...folk scene and for people...that love folk music, because you can't deny that...there's an absence of content about Black characters [in the] canon of traditional songs that we have.[28]

Morrison's own heritage and biography are woven into the story she is trying to document. She was born in Birmingham, in the heart of England; her mother is from Jamaica and her father is Scottish, from the Outer Hebrides. She first heard folk music on the radio when she was eight or nine years old:

> I just happened to be playing at home, and my parents were listening to Radio 4 or something. And it's just lots of adults talking. It was so boring. And...I change it out until this extraordinary unaccompanied human voice began to sing. And I'd never heard anything like it at all. It was like a voice from another world. It was like a voice from another era. And I've never forgotten that voice. And as an adult, I found out...it was Shirley Collins, singing "Our Captain Pride."...You know, music that you love gives you the shivers all over, and there are few feelings as delicious as that, but that's what I got. I got absolutely the shivers.[29]

She went to her first folk club at the age of seventeen, with her father. She went on to become active in the Midlands music scene, educating herself in the tradition and the music. She comments, "All my experiences in folk clubs have been overwhelmingly positive. As anybody in the folk scene will tell, you know, they're incredibly friendly places. They're lovely. You get the lovely feeling of being amongst people who welcome you because you love the same thing that they love. At the same time, you are always the only Black person in the room."[30] Strikingly, folk clubs were welcoming and receptive to the stories of Black life in Morrison's songs, and yet at the same time, folk clubs and the folk scene more generally remain enduringly white, largely in social spaces of the middle-aged and middle class.

Rather than using cultural fusion or combination in her music, Morrison introduces issues of slavery, empire, and belonging into traditional folk song and into the heart of the folk community through collaborating with Eliza Carthy and Martin Carthy, in many ways the first family of English folk music.

Morrison summed up the broader cultural and political purpose of this music:

> I wanted to give voice to some of these souls of Black Britain, and of Black Albion, and . . . all of these silenced ancestors deserve to have their voices heard. I also very much want to present within the context of the folk scene or the folk community, to present an imaginary past, which includes Black people. Because Black people really were there. . . . these ghosts of old Black Albion . . . will receive the kind of hospitality that . . . has been denied to them all this time. . . . These people belonged in the land and belonged to the woods, to the trees, to the rural places, to the coastal spaces to the mountains. . . . I want these ghosts to populate those places.[31]

"Wobbles on Cobbles": Hak Baker and G-Folk

The music of Hak Baker (see fig. 2) and his brand of G-folk is less concerned with recovering the traces of the past and more with documenting unheard voices today. Baker's music is born of his experiences of growing up in dockland London, of life on the Isle of Dogs, a poor district of East London on the inside meander of the River Thames just across the Thames from where Bert Lloyd was writing forty years earlier. Baker's songs portrays this corner of the capital in striking realism, shedding light on its beauty and shadows. Although his music defies categorization, and he cares little for conventional musical labels, we argue it contains a twenty-first-century folk sensibility that attends to and documents remarkable lives that are otherwise unremarked on.

Baker was born in Luton on June 16, 1991, and moved to the Isle of Dogs when he was a young child. We first met him in the summer of 2021, following the success of his song "Wobbles on Cobbles," released in 2020.[32] It is an urban fable that documents run-ins with the police and East End struggles to make ends meet, but it also captures the stifling feel of life in COVID-19 London. He explained in an interview,

Figure 2. Hak Baker, 2023. Photograph by Nickii Kane.

Most my music comes to me like streams of consciousness, like, I feel like it shouldn't be that hard to write a song, if we're telling the truth or if you're like, you know, just, admitting how you feel. So, those words . . . that's just how I felt. And, I had the guitar riff, I was playing it at home, "doo doo doo." And then I brought my mate Renee to come over because I felt like it needs some keys to make it like. . . . I said, "come over, I'm writing the song," and she came. And, it was like a nursery rhyme, and I showed her, and then she just went "doo doo doo doo." And then, uh, yeah, just started writing it. And it just came out in a day.[33]

There are so many songs about London life. They are often the voice of hope and the demand for something else, and they are very often, like Hak Baker's music, situated in a particular part of the city. Like the best folk songs, they have the power to make us notice and feel that which is all around us yet unnoticed, and to create a deep connection to place. Baker continued,

> Where we are from and . . . that generation that we managed to just hold on to hard work and a bit of struggle, it comes synonymous, and you'd have to help each other and you'd smile, and your front door would be open, and I really learned that culture because I am of Caribbean culture. My mum was really strict—don't talk to strangers, don't do this, don't do that, you know, be in at fucking six o'clock or seven o'clock, and I was like this stuff don't make no sense. . . . And when I went to secondary school, and I was really . . . introduced to exactly where I was actually living, that was it. That was it, mate. I just fell in love with everything about where I'm from.[34]

Songs help us read the "signs in the street," borrowing from the ideas of the late great urbanist Marshall Berman.[35] The Isle of Dogs that Baker paints in his songs is not the portrayal of a racist area. This part of London did become notorious in 1993, when Derek Beakon won a by-election for the BNP, the first elected representative for the party. Baker's mum felt those associations and, in a way, wanted to protect her son from this place.

> She used to say things like, "There are racist people out there." If we didn't get that, then all of us would fight—not just the black boys, we'd all fight, do you know what I mean? And that's what the Isle of Dogs was, it was just like a melting pot of like, a lot of ethnicities and people. And, you know, I feel like, even though it was like an old-school, like an NF [National Front] place, my mum used to say, . . . I didn't really feel that because for me, . . . it was like this old-school dockers town, and it was about that, you know, hard work and . . . being at the bottom of the barrel. You know . . . no one used to come to the Isle of Dogs. I felt like we was all on this little island, and we all went to the same schools and that, and then that was it. I remember like my best friend Tia—who I'm still best friends with now, but she's white—. . . there's never been an instance where we've ever felt like anything [racial]. And like, she was . . . literally probably one of my first friends ever. My friend Emrah, Turkish boy, used to go in his house all the time. My friend Asher, Asian boy, still remember them. Phi, Chinese Boy. You know, Anton—like Black boys that lived on my road. I just . . . didn't ever felt it. I see it and I experienced it—don't get it twisted, definitely. But not really, not on the Isle of Dogs.[36]

Music was a central part of his life, but it came from a huge range of sources. "Funny old story, the old music, you know," Baker shares with his characteristic charm. His father is from Grenada and converted to Islam, and the family lived initially in Luton, a small town thirty miles northwest of London, where he ran a reggae sound system at Luton Carnival. His mum listened to reggae, revival, and dancehall: "Everything in the car, Sunday morning, cooking, so I was very always open to like reggae and stuff. . . . And, so I was obsessed with Bob Marley from a young age. He's always been the champion of all champs."[37]

Then in primary school his life and relationship to music took a surprising turn. Baker remembers,

> I won a singing competition. Just normal, just singing. I come home, and then . . . my mum said . . . you're going to start going to this place every Friday to sing. I didn't know what it was, but then I came to realize it was Southwark Cathedral choir in London Bridge. Then one day they was like to me, "On Sunday, you got to come in, you're going to be inducted as a chorister." I was like, "What?" Yeah, so then, like, I came home, I went to my mum, like, "Mum, these people saying I'm gonna be like a choir boy." She was like, "Yeah, man" [laughs].[38]

The huge social distance between working-class life on the Isle of Dogs and the rarefied elite atmosphere of the Southwark Cathedral choir was hard to navigate. Southwark Cathedral is the same place Nadia Cattouse sang about in her ballad "Bermondsey," at the Edinburgh Festival in 1969. All these stories relate to common threads in this cultural tapestry. "I did that for four years and hated it, but like, you know, I learned a lot from there, like, musically," Baker reflected. "I really liked the ambience and how it reverberated around in the altar and the church and stuff." The English choral tradition, particularly the compositions of Ralph Vaughan Williams, is one of the main areas where classical music incorporates folk song. We feel sure that, if Vaughan Williams were alive today, he would be tuning his ears to new voices like Baker's. He loved the music, and it has left a lasting impression and influence on him: "All that old Latin shit and all the old English shit, all of that shit. To the max, you know. We went [to] Westminster Abbey, sung there, sung at Oxford, sung in front of the Queen—never used to bow, mum said don't bow."[39] The world of the choristers and the choir, inflected with middle-class dispositions and cultural capital, proved just too uncomfortable a place, and Baker faked his voice breaking as an excuse to leave the choir. His experience is a salutary one in terms of the continuing and exclusive aspects of classical music and its relationship to the English folk tradition.[40]

Leaving the choir, he became involved in the grime music genre and more contemporary forms of urban music and lyricism. He became a member of the BOMB Squad E14 and had success as an MC and rapper.[41] The group had success and exposure on Channel U (later known as Channel AKA), a British digital satellite TV music channel that focused on the UK grime scene, which enabled MCs to submit their music via handheld digital cameras or phones. Channel U provided a platform for such British grime acts as Tinchy Stryder, Tinie Tempah, Dizzee Rascal, Chip, Wretch 32, Devlin, Giggs, Skepta, and N-Dubz to break through to wider audiences. In a way, these self-regulated and managed platforms echo the qualities of folk music's song school or singaround, which has led some to conclude that rap and drill are the "folk music of today."[42] However, as Tilman Schwarze and Lambros Fatsis have pointed out, it is through digital platforms like YouTube that grime and drill have been criminalized and subjected to police scrutiny.[43] Baker, like so many young Black working-class men, fell foul of the law and served a prison sentence. Baker picks up his story:

> We'd...MC everywhere, go to all the youth clubs and stuff. And then, um,...I went to jail. And then...one day, the governor come in the wing, and it was like, "Ah, there's guitar lessons going. Do you want some?" And, I was like, "I'll give that go," because, by that time, just before I went in that time...me and my mate Josh was listening to like Kings of Leon and like all kind of guitar-based music, sometimes. I was really into it. So, I was like, "Yeah, I'll give that a go." And, my name got picked out of a hat. And that was it—I started playing the guitar.[44]

Baker's voice, and the meter of his songs, is in part trained through singing in church, but he found the guitar, his main harmonic instrument, while in prison. Out of prison, his friends encouraged him to apply for a spot in the Levi's Music Project. He was selected and joined ten other aspiring musicians who were trained in how to become better musicians. This was achieved by exposing them to good music mentors, equipment, and quality studio sessions. The project was headed by grime superstar Skepta.

Baker's career began in May 2017 with the release of his debut single "7AM." His first album, *Babylon*, released in 2019, solidified Baker's spot as a leading UK singer, one that knew how to make entertaining music. His storytelling ability as a songwriter, which highlights the challenges of being young, Black, and working class, is what distinguishes his music. Baker's songs make "kind villains" and "nighttime stars" the heroes of city life. The music also confronts the experience of pain and darkness and the strength that comes from facing up to it.

The influence of his training in the choir and the English folk tradition is there in the form, rhythm, and harmonic structure of so much of his music. Baker explained,

> It comes in different ways...like the way that I generally like to play in 3/4 [time]. And there's a lot of 3/4 stuff in the church. And I just think there's a lot more space there to be silky, you know what I mean. I was like, my mum was always working, so she'd be late. I really can't believe I can't remember this geezer's name, but one of the temporary choir masters that came there, but he was just a young guy, normal guy....He could see that I didn't fit.

And he would spend time with me after trying to teach me stuff and playing the piano with me, and I can't believe I forgot his name.

Yeah, but like now I use those harmonies all the time. Like now, . . . I've been encouraged to sing a lot more and that's the singing technique that comes actually the most natural to me—that's what I've been trained in. So, I use it a lot. People are kind of shocked when I sing.[45]

There is a sense of loss and grieving in many of Hak Baker's songs. "Grief Eyes," released in 2019 and included in his album *Babylon*, is an incredibly touching ballad of lives cut short prematurely in London. It also honors the everyday routine kindness of city life, particularly in the face of hardship. Baker explained,

Richie was like an old-school white geezer that, uhm, first geezer put the gloves on my brother, who went on to be a boxer and, you know, wouldn't take a penny for the kids for the boxing lessons. Like, even when I was getting in trouble, he would like grab me and like say come work for me at, like, the scrap yards and stocking tires and stuff. When I used to go court, he used to go court and talk absolute bollocks and say that I was a this and that and that I had been . . . helping in a youth club and whatnot—a kind villain.[46]

In describing a kind villain who was more a local hero, Baker sets the moral complexity of multicultural cities to music. Richie suffered with multiple sclerosis, and the video for "Grief Eyes" shows him and the world from which the song emerges.[47] "I look in his eyes, and I see a sadness that's so dark, so deep that hopefully one day he talks to someone about it. We talk sometimes when they're drunk. Yeah. So that's what that song is about. My mate Eugene [who is mentioned in the lyric of the song] is in jail for twelve and a half years for some stupidness, but yeah, just a look around, innit."[48] The best folk songs make the listeners inhabit the world of the characters within them. Baker's music is like this, and he makes us confront the joy and pain and particularly the complexity of the lives of young men. Baker makes us walk in their shoes as they try to make their way in an unforgiving city that judges them harshly. He calls the songs that document these lives "guv'nor folk," or G-folk. It is a definition of folk music Bert Lloyd likely would have approved of: songs and ballads of those at the edges of or who are cast out from the wealth and privilege of London as a global city.

In May 2022 Baker announced the Bricks and Mor-Tour of public houses in eight UK cities, from Glasgow to London. The British pub is an iconic place both within working-class life and in the folk imagination. The tour was about reconnection with his audience, having a pint with the punters before and after the show. He reflected, "They can see I'm one of them. . . . Playing to people in pubs really just puts your gravity boots right back on. . . . The old school pub energy could save the world. A stranger walks in, the locals see there's a stranger walking in, eye him up a couple times, then go, 'Do you want a drink, mate?' Things as simple as that."[49] For him this offers a space for public life and cultural value, and the importance of music performance in small venues has also been acknowledged by academic researchers as having the capacity to create social bridges.[50]

As Monique Charles and Mary Gani argue, the notion of "urban music" has become a straitjacket for young Black creativity in the United Kingdom.[51] This point is developed by Anamik Saha in *Race and the Cultural Industries*, where he documents the pervasiveness of very conservative and restrictive racialized formats within the record industry.[52] Baker has been victim of this kind of stereotyping from journalists because his artistry breaks these molds:

> I've had ridiculous things people say to me. "How does it feel to be a black man playing guitar?" And, I said, "Are you fucking dumb. Like do you . . . how can you even be a journalist and ask me that question?" They think like, because of how I look I am going to be a rapper or something. When I open my mouth, when I've got this East London dialect, or it is like the London dialect, you know, . . . from the Isle of Dogs, being from 'round here, and that, that'll just throw them off anyway. And then the guitar, like . . . we've been the stars of the guitar since forever, you know, Africa, then the blues, like forever, we've been smashing the guitar. So . . . how can you ask me that? It is silly. But at the same time, I feel like I've always been widening the boundaries of what any young person in London is deemed to be doing anyway.[53]

Baker's 2023 album *Worlds End FM* draws from a range of influences, including folk, punk, and reggae. On the eve of its launch we spoke to Baker about his music, influences, and Englishness. Describing the music of his latest album, Baker says, "Tunes are big, tunes are massive, tunes are fun, tunes are thought provoking, tunes are sad. The lyrics are super current."[54] Tunes like "Windrush Baby" and "Telephones 4 Eyes" are folk songs capturing the darker and complicated sides of life in London today, from the addiction to smartphones and the digital capture of the self via LinkedIn to the withering of street life:

> Now the love is gone from the street
> Can you not see it?
> Just leave it to the Windrush baby
> It's time we reclaim it, so meet me at the street at night[55]

World's End FM's sound draws from influences from punk to psychedelic rock. It plays with the idea of serving as a "pirate radio station" that broadcasts songs but also takes calls from friends and fans that phone in with dispatches from a world that's falling apart. He explained, "I feel like that's the disarray that I'm seeing in the world right now. And I feel like the only way for me to really get that out is through fucking punk poetry."[56]

If we think of folk music as Bert Lloyd did, as a living tradition passed on orally, telling stories, and painting the life of ordinary people, then this is the sound of twenty-first-century English folk music. This is a sense of folk music that abandons any unifying essence of what Paul Gilroy calls "ethnic absolutism" based on race or a melancholic attachment to empire.[57] It revels in an impure Englishness that is in a constant unfinished state of emergence. Baker continued,

> I feel like a lot of people have got this thing about what it is to be English, but England is a place that has been conquered many, many, many times, until they did the conquering like in the 1800s and 1900s. So essentially, we're a mongrel society. And if people look into that, then they should understand that we should just all be mobbing about being together because we are a massive mongrel society, but people don't know their own history. So, no wonder there's loads of us in all different colors, shapes, and sizes, because we're a mongrel society.[58]

Conclusions

As Billy Bragg championed in his song "English, Half English," there is often a glossing of this ancient mixing in discussions of English cultural identity. Folk music is often derided or ridiculed for its quintessential rural associations, as it is in essence a music of the people. However, it has become a particularly interesting, contested field that is struggling with the meanings of Englishness and Britishness. Losing its narrow associations with whiteness and nationalism has given English folk music new vitality in ways that its early exponents like Bert Lloyd would probably approve of. Toward the end of his life, Lloyd admitted to filmmaker Barry Gavin that the "definition of what is folk and is not, is beyond me.... I wouldn't attempt to define.... After all it's like the difference between day and night.... We know the difference between day and night, but you can't say exactly where the break occurs and which is darkness on one side and light on the other."[59] This leaves open the possibility for a boundless conception of what folk music is, as an expressive medium for the human voice in song, carrying all its infinite variety.

In this discussion we have tried partly to recover the "shadow history" of Black folk in English folk.[60] More than that, though, we want to break the legacy of racialized thinking in the idea of "the folk" in folk music. This is what Morrison's music does so effectively, restoring the ghostly traces of Black Albion within the medium of traditional folk song that proliferates rather than homogenizes cultural, local, class, racial, and sexual differences.

Artist Ben Edge commented to *Rolling Stone Magazine* in May 2023,

> I think young people are engaging with folklore because it's a way of opposing nationalism in an organic grassroots way.... As with the victims of the Empire, indigenous culture was taken away from the peasants when they were pushed off their land, so there's a need to reconnect. Folk has always been the culture of the people, which has nothing to do with the upper classes and their stately homes. Coming at it from that angle, folk culture can create a more inclusive Britain today.[61]

Edge points to the possibility of undermining the association between English folk music and whiteness and creating an open folk tradition from below.

We have examined the music of Angeline Morrison and Hak Baker and argued that these artists of color are good examples of what twenty-first-century English folk song sounds like in its various sonic textures, references, and audiences. Embracing these Black voices as part of the contemporary folk tradition also opens folk up to include other marginalized and excluded groups. For example, many of the members of the Shovel Dance Collective folk band identify as nonbinary; consider also LGBTQIA+ sea shanty choir Seaweed in the Fruit Locker, which takes sea shanties, often depicting problematic historical perspectives, and updates them to make them relevant to contemporary LGBTQIA+ themes. Folk is a music of the people, of the streets, and of communities. It is a varied and evolving musical form that has the capacity and potential to include and champion the wide variety of what it means to be English in the United Kingdom today—we just need to open our eyes and ears and hearts to what Stuart Hall called "our mongrel selves."[62] However, this contains a risk of being trapped in what Tavia Nyong'o calls the "amalgamation waltz," which has "been repeatedly enlisted to envision utopian and dystopian scenarios."[63] Here, the idea of "mongrel" can be shared by white phobias of miscegenation as well as the multicultural dreams of transcendent cultural mixture. We suggest a different kind of understanding of folk song and dance that isn't tied to such a racial logic, one that can break with its genealogy to document the stories of English folk, past and present, in all their irreducible varieties of joy, pain, pride, damage, confinement, and freedom.

Les Back is professor of sociology at the University of Glasgow. He is also a journalist, broadcaster, and musician. His books include *Migrant City* (with Shamser Sinha, Charlynne Bryan, Vlad Baraka, and Mardoche Yembi; 2018) and *The Unfinished Politics of Race* (with Michael Keith, John Solomos, and Kalbir Shukra; 2022).

Stevie Back is an urban farmer, life coach, copy editor, and yoga teacher from London. Music is one of her many loves, and she is a regular attendee of her local weekly folk music sing-around, a hopeful, diverse (and predominantly young) space where people gather and share traditional songs from around the world.

Notes

We thank Angeline Morrison and Hak Baker for speaking to us and enriching this article with their reflections on their music making. We also thank Paul Gilroy, Vron Ware, Elspeth Merry, Tariq Jazeel, and Tom Western for their tips and insights, which enriched our thinking, and the two anonymous reviewers, who encouraged us to think harder about what is at stake in thinking critically about English folk music historically and today.

A Spotify playlist titled "Black Folk in English Folk" accompanies this article and provides a sound track for almost all the music mentioned: https://open.spotify.com /playlist/6r1gFWdAo5cNhvVIGn2uoP?si=yTZPs7R1QaiLu1cSW9bn0A&nd=1.

1. Lloyd, *Singing Englishman*, 10–11.
2. Gregory, "A. L. Lloyd and the English Folk Song Revival."
3. Sharp, *English Folksong*; Harker, *Fakesong*.
4. Lloyd, *Folk Song in England*.
5. For Lloyd's definition of folk music, see Arthur, *Bert*, 245–46.
6. See Wilks, "Angeline Morrison."
7. Gilroy, "English Folk Tradition and the Choice of Ancestors," 9.
8. Gilroy, "English Folk Tradition and the Choice of Ancestors," 18.
9. Lloyd, *Life in the Elephant*, 10.
10. Williams, *Marxism and Literature*, 126.
11. Morton, *People's History of England*.
12. Morton, "A. L. Lloyd: A Personal Memoir."
13. Lloyd, *Folk Song in England*, 71.
14. Back, "Coughing Up Fire"; Henry, *What the Deejay Said*.
15. Cecil Sharp diary, December 8, 1918, p. 345, Vaughan Williams Memorial Library, Cecil Sharp House, London, https://www.vwml.org/record/SharpDiary 1918/1918/p345.
16. Gilroy, "Belonging and Un-belonging in the English Countryside."
17. This song is not available on Spotify and is not included in the playlist that accompanies this article, but it is available here: https://www.youtube.com/watch ?v=8L-rI395_nw.
18. Sparrow, "How Paul Robeson Found His Political Voice."
19. Robeson, *Here I Stand*, 115.
20. Williamson, "British Folk Comes of Asia," 40.
21. See Ware and Back, *Out of Whiteness*, chap. 4.
22. Koch, "Osama Van Halen and the 50 Cent Dictator."
23. Lucas, "Imagined Folk of England," 7.
24. Chipping, *Folk against Fascism* liner notes.

25. Lucas, "Imagined Folk of England," 16.

26. Du Bois, *Souls of Black Folk*.

27. Angeline Morrison, telephone interview by Les Back, January 11, 2024.

28. Morrison interview.

29. Morrison interview.

30. Morrison interview.

31. Morrison interview.

32. *Cobbles* is a London Cockney phrase for streets that in the days before tarmacadam were cobbled, uneven, and not smooth. To "wobble on the cobbles" is to lose your footing or balance, to stumble or be unstable.

33. Hakeem Baker, interview by Les Back, London, August 20, 2021.

34. Baker interview, 2021.

35. Berman, "Signs in the Street," 123.

36. Baker interview, 2021.

37. Baker interview, 2021.

38. Baker interview, 2021.

39. Baker interview, 2021.

40. Bull, "Challenging Classical Music's Genre Conventions."

41. BarcodesTV, "Bomb Squad."

42. See Robertson, "Why Rap and Drill Are the Folk Music of Today."

43. Schwarze and Fatsis, "Copping the Blame."

44. Baker interview, 2021.

45. Baker interview, 2021.

46. Baker interview, 2021.

47. Robson-Scott, "Hak Baker."

48. Baker interview, 2021.

49. Hakeem Baker, interview by Les Back, London, February 7, 2023.

50. Behr, Brennan, and Cloonan, *Cultural Value of Live Music*, 3.

51. Charles and Gani, *Black Music in Britain in the Twenty-First Century*, 1–8.

52. Saha, *Race and the Cultural Industries*.

53. Baker interview, 2021.

54. Baker interview, 2023.

55. "Windrush Baby" lyrics, Hak Baker *World's FM* album liner notes, Hak Attack Records/AWAL, HAKO12V, 2023.

56. Baker interview, 2023.

57. Gilroy, *There Ain't No Black in the Union Jack*, 61; Gilroy, *After Empire*.

58. Baker interview, 2023.

59. Quoted in Gavin, *Singing Englishman*.

60. Gilroy, "Belonging and Un-belonging in the English Countryside."

61. Edge, quoted in Goto, "Less Pale Male and Stale."

62. Hall, "Our Mongrel Selves."

63. Nyong'o, *Amalgamation Waltz*, 175.

References

Arthur, David. *Bert: The Life and Times of A. L. Lloyd*. London: Pluto Press, 2012.

Back, Les. "Coughing Up Fire: Sound Systems and Cultural Politics in South East London." *New Formations* 8 (1988): 141–52.

BarcodesTV. "Bomb Squad (B.O.M.B) Exclusive Grime Freestyle (Barcodes DVD Rare Footage)." YouTube, September 27, 2018. https://www.youtube.com/watch?v=I3UWXkgJ7v0.

Behr, Adam, Matthew Brennan, and Martin Cloonan. *The Cultural Value of Live Music from the Pub to the Stadium: Getting beyond the Numbers.* Live Music Exchange, 2014. https://www.pure.ed.ac.uk/ws/portalfiles/portal/16546029/The_Cultural_Value_of_Live_Music_From_Pub_to_Stadium.pdf.

Berman, Marshall. "The Signs in the Street: A Response to Perry Anderson." *New Left Review* 144 (March/April 1984): 114–23.

Bull, Anna. "Challenging Classical Music's Genre Conventions: Findings from a Project on Youth Voice in Instrumental Education." *Journal of the Royal Musical Association*, forthcoming.

Charles, Monique, and Mary Gani. *Black Music in Britain in the Twenty-First Century.* Liverpool: Liverpool University Press. 2023.

Chipping, Tim. *Folk against Fascism* liner notes. FAF1CD, 2010.

Du Bois, William Edward Burghardt. *The Souls of Black Folk.* 1903; repr., New York: Bantam, 1989.

Gavin, Barry, dir. *The Singing Englishman: A Portrait of A. L. Lloyd.* 1983. https://www.youtube.com/watch?v=4k3Amr_qE5o.

Gilroy, Paul. *After Empire: Melancholia or Convivial Culture?* London: Routledge, 2004.

Gilroy, Paul. "Belonging and Un-belonging in the English Countryside." Lecture given at the Claiming Belonging through Looking and Listening symposium, Milton Keynes Gallery, May 2, 2019. https://www.youtube.com/watch?v=WZbgchdC2z0.

Gilroy, Paul. "English Folk Tradition and the Choice of Ancestors." *Folklore* 135, no. 3 (2024): 313–26.

Gilroy, Paul. *There Ain't No Black in the Union Jack: The Cultural Politics of Race and Nation.* London: Hutchinson. 1987.

Goto, Zoey. "Less Pale Male and Stale—How British Folk Is Getting a Remix." *Rolling Stone*, May 15, 2023. https://www.rollingstone.co.uk/music/features/less-pale-male-and-stale-how-british-folk-is-getting-a-remix-29180/.

Gregory, E. David. "A. L. Lloyd and the English Folk Song Revival, 1934–44." *Canadian Journal for Traditional Music* 25 (1997): 14–28.

Gregory, E. David. "A. L. Lloyd: A Personal Memoir." *Englisch-Amerikanische Studien: Zeitschrift für Unetrricht, Wissenschaft und Politik* 4 (1984): 688–90.

Hall, Stuart. "Our Mongrel Selves: The Raymond Williams Memorial Lecture." In "Borderlands," supplement to *New Statesman* 5, no. 207 (1992): 6–8.

Harker, David. *Fakesong: The Manufacture of British Folksong 1700 to the Present Day.* Milton Keynes, UK: Open University Press. 1985.

Henry, William. *What the Deejay Said: A Critique from the Street!* London: Learning By Choice, 2006.

Koch, Christian. "Osama Van Halen and the 50 Cent Dictator." *Guardian*, October 16, 2010. https://www.theguardian.com/music/2010/jan/16/osama-ipod-guardian-guide.

Lloyd, Albert Lancaster. *Folk Song in England.* London: Lawrence and Wishart, 1967.

Lloyd, Albert Lancaster. "Life in the Elephant." *Picture Post* 42, no. 2 (1949): 10–17.

Lloyd, Albert Lancaster. *The Singing Englishman: An Introduction to Folksong.* London: Workers' Music Association, 1944.

Lucas, Caroline. "The Imagined Folk of England: Whiteness, Folk Music, and Fascism." *Critical Race and Whiteness Studies* 9, no. 1 (2013): 1–19.

Morton, Arthur Leslie. *A People's History of England.* London: V. Gollancz, 1938.

Nyong'o, Tavia. *The Amalgamation Waltz: Race, Performance, and the Ruse of Memory.* Minneapolis: University of Minnesota Press, 2009.

Robeson, Paul. *Here I Stand.* Boston: Beacon, 1958.

Robertson, Mark. "Why Rap and Drill Are the Folk Music of Today." *Guardian,* August 4, 2023. https://www.theguardian.com/music/2023/aug/04/why-rap-and -drill-are-the-folk-music-of-today.

Robson-Scott, Will. "Hak Baker—Grief Eyes." Vimeo, September 18, 2019. https:// vimeo.com/360842704.

Saha, Anamik. *Race and the Cultural Industries.* Malden, MA: Polity Press, 2017.

Schwarze, Tilman, and Lambros Fatsis. "Copping the Blame: The Role of You-Tube Videos in the Criminalisation of UK Drill Music." *Popular Music* 41, no. 4 (2022): 463–80.

Sharp, Cecil J. *English Folksong: Some Conclusions.* London: Simpkin and Novello; Taunton, UK: Barnicott and Pearce, 1907.

Sparrow, Jeff. "How Paul Robeson Found His Political Voice in the Welsh Valleys." *Guardian,* July 2, 2017. https://www.theguardian.com/books/2017/jul/02 /how-paul-robeson-found-political-voice-in-welsh-valleys.

Ware, Vron, and Les Back. *Out of Whiteness: Color, Politics, and Culture.* Chicago: University of Chicago Press, 2002.

Wilks, Jon. "Angeline Morrison—The *Sorrow Songs* Interview." *Tradfolk,* December 16, 2021. https://tradfolk.co/music/music-interviews/angeline-morrison-sorrow -songs/.

Williams, Raymond. *Marxism and Literature.* Oxford: Oxford University Press, 1977.

Williamson, Nigel. "British Folk Comes of Asia." *London Times,* July 2, 1999.

Yale UNIVERSITY PRESS

Empty Spaces
Jordan Abel

Storylife
On Epic, Narrative, and Living Things
Joel P. Christensen

No One Will Know You Tomorrow
Selected Poems, 2014-2024
Najwan Darwish
The Margellos World Republic of Letters Series

As If Human
Ethics and Artificial Intelligence
Nigel Shadbolt and Roger Hampson

The Business of Killing Indians
Scalp Warfare and the Violent Conquest of North America
William S. Kiser
The Lamar Series in Western History

Tattoos
The Untold History of a Modern Art
Matt Lodder

BLACK LIVES SERIES

John Lewis
In Search of the Beloved Community
Raymond Arsenault

James Baldwin
The Life Album
Magdalena J. Zaborowska

What Nails It
Greil Marcus
Why I Write Series

Where We Stand
Djamila Ribeiro
The Margellos World Republic of Letters Series

Provincials
Postcards from the Peripheries
Sumana Roy

The Plunder of Black America
How the Racial Wealth Gap Was Made
Calvin Schermerhorn

The Performer
Art, Life, Politics
Richard Sennett

There Is a Deep Brooding in Arkansas
The Rape Trials That Sustained Jim Crow, and the People Who Fought It, from Thurgood Marshall to Maya Angelou
Scott W. Stern
Yale Law Library Series in Legal History and Reference

Robert Wedderburn
British Insurrectionary, Jamaican Abolitionist
Ryan Hanley

Bigger
A Literary Life
Trudier Harris

AVAILABLE IN PAPERBACK

On Marriage
Devorah Baum

The Rediscovery of America
Native Peoples and the Unmaking of U.S. History
Ned Blackhawk
The Henry Roe Cloud Series on American Indians and Modernity

Yale and Slavery
A History
David W. Blight

The Abduction of Betty and Barney Hill
Alien Encounters, Civil Rights, and the New Age in America
Matthew Bowman

They Flew
A History of the Impossible
Carlos M. N. Eire

The Fine Art of Literary Fist-Fighting
How a Bunch of Rabble-Rousers, Outsiders, and Ne'er-do-wells Concocted Creative Nonfiction
Lee Gutkind

Becoming Irish American
The Making and Remaking of a People from Roanoke to JFK
Timothy J. Meagher

Roe
The History of a National Obsession
Mary Ziegler

yalebooks.com